Concrete's Environmental Impact and the Alternatives

Concrete's Environmental Impact and the Alternatives

Authors
Tim Chapman
Zarish Jawad
Sudipta Samadder
Sarah Mansoor
David Henneberg
Syed Rizvi
Chitrini Tandon
Maggie Wang
Kanish Baskaran
Daniel Gurin
Lea Touliopoulos

Editor
Catherine Mardon

PRESS

Copyright © 2023 by Austin Mardon

All rights reserved. This book or any portion thereof may not be reproduced or used in any manner whatsoever without the express written permission of the publisher except for the use of brief quotations in a book review or scholarly journal.

First Printing: 2023

Typeset and Cover Design by Josh Harnack

ISBN: 978-1-77369-900-4
eBook ISBN: 978-1-77369-901-1

Golden Meteorite Press
103 11919 82 St NW
Edmonton, AB T5B 2W3
www.goldenmeteoritepress.com

Table of Contents

Foreword 7
Chapter 1: Introduction 9
Chapter 2: Ferrock 24
Chapter 3: Ashcrete 34
Chapter 4: Timbercrete 45
Chapter 5: Aircrete 56
Chapter 6: Hempcrete 66
Chapter 7: Recycled Plastic 75
Chapter 8: Mycelium 85
Chapter 9: Bamboo.. 95
Chapter 10: Popular Projects Using Alternative Materials................ 107
Chapter 11 The Timeline of Concrete 120

Foreword

Concrete, the unsung hero of construction. This strong and versatile substance has been used for centuries to build everything from homes to bridges to skyscrapers. It's tough, durable, and can be molded into any shape you can imagine. However, From its production to its disposal, concrete contributes significantly to greenhouse gas emissions, energy consumption, and the depletion of natural resources. However, there is growing awareness of the need to reduce our carbon footprint and find more sustainable alternatives. In 11 short chapters, this book aims to explore the environmental impact of concrete, one of the most widely used building materials in the world. Join us as we examine the alternatives to concrete, from traditional materials such as bamboo, to new and innovative materials such as aircrete, mycelium, recycled plastics and more. We also look at the pros and cons of each alternative, and the factors that influence their adoption. The goal of this book is to provide a comprehensive overview of the environmental impact of concrete and the alternatives available, so that we can make informed decisions about how to build a more sustainable future.

In Chapter 1, we begin with an introduction to concrete itself. This chapter touches on the historic use of concrete and its importance, from Roman concrete to modern cement, the importance of this compound can not be disputed. This chapter finishes with a brief discussion of alternatives to concrete and how they can be used to build the future. In Chapter 2, we begin our in depth discussion of concrete alternatives, beginning with Ferrock. Formed primarily out of waste steel or iron, this compound functions as both a waste management tool that could be the next step in construction. The reader will learn about its history, how it is made, and more. Chapter 3 touches on Ashcrete. Here, the reader will discover its green impact while learning about the environmentally-friendly construction material. This chapter includes discussions of fly ash, borate, bottom ash and more. In Chapter 4, we take a look at Timbercrete, an alternative form of concrete formed by saw dust. This chapter focuses on Timbercrete's suitability for a wide range of building projects, including homes and residential buildings, while also serving as a carbon sink. In Chapter 5, we continue our green alternatives with Aircrete. This chapter will focus on the history of Aircrete, its potential applications in the future of construction, as well as the environmental and practical impact it will have as it starts to replace regular concrete (especially in North America). Chapter 6 sees us visit Hempcrete. We take a look into the attractive properties of Hempcrete, including its properties of carbon-sequestering, higher biomass production and its high levels of insulation. In Chapter 7, we discuss the role of recycled plastic as an alternative to concrete. This chapter dives into the utilization of plastic waste in cement, while also identifying a number of scientific advances and studies. In Chapter 8, we take a look at Mycelium. This chapter identifies key factors that the largest living organism on earth possess when it comes to finding a viable alternative to concrete. Could building with mycelium-based composite materials be the future? In Chapter 9, we address the possibility of utilizing bamboo as an alternative. Could one of the world's fastest growing plants be the key to faster growing cities? This chapter dives into the structural composition of bamboo, how it is harvested, and how its use could be a viable alternative to conventional building materials among other topics. In Chapter 10, popular projects that have utilized alternative materials are discussed. Learn about plastic bottle roads, the use of RPM asphalt mixes, plastic bridges, and other feats of constructive innovation from all corners of the world. Finally, in Chapter 11, we address the timeline of concrete and how this revolutionary building material gained its control over the modern building world.

Chapter 1: Introduction
by Tim Chapman

Throughout world history there are few examples of technologies that have elevated the societies that utilized them more so than the invention of concrete. An everyday item that the majority of the population scarcely thinks about let alone has an in-depth knowledge of its history and more importantly its environmental impacts. The term concrete does not refer to a single solitary item but rather concrete is a composite material comprised of a mixture of fine and coarse aggregate (sand, gravel, etc) and a fluid cement (burnt limestone known as quicklime) that hardens through a chemical process over a period of time (Gagg, 2014).

Historians actively debate the first use of concrete with records being found across the globe at various points of time, e.g. Chinese records first uses of concrete in 2000 BCE, Mexican records dating back to 1000 BCE (Jahren & Sui., 2017). The first societies to use concrete in the same fashion we do today is undoubtedly the Romans who began using the substance in roughly 300 BCE with naturally formed concrete being dated back around twelve million years ago (Brewer, n.d.). Roman concrete consisted of quicklime, pozzolana, which is a particular type of ash that chemically hardens when mixed with quicklime and water, and an aggregate of Pumice stone (volcanic rock).

The invention of Roman concrete was the catalyst for the Roman architecture revolution that saw the Roman architects free themselves from the hands of stone/brick construction. Resulting in the construction of commonly recognized historical sites such as the Pantheon and the infamous Roman Colosseum which are both still standing thanks to the revolutionary Roman concrete. Despite the lengthy recognized history of concrete that we have today is in contrast to the Romans historical records who recognize themselves as the first inventors of concrete (Robertson, 1969). With the decline of the Roman empire beginning in 476 AD, the commonly used combination of quicklime, pozzolana and pumice was dramatically reduced. Historians believe that a drop in kiln temperatures used for turning lime into quicklime, a lack of pozzolana and poor technique were all contributing factors to the reduction of Roman concrete after the fall of the Roman Empire (Diamond, 2011).

The globe would not see concrete that was equal to Roman concrete for roughly 900 years with this gap filled with increasingly stronger mortars as the replacement (Harrison et al., 2019 pp. 341-342). By the seventeenth century concrete makers were adding pozzolana into their mixtures, making their finished work closer to that of the Romans which has lasted the test of time. Within the next hundred years, concrete makers and bricklayers had mastered the use of hydraulic lime which is cured with hydration rather than with carbon dioxide, making a much stronger product.

Modern cement like we know it today was not invented until 1824 when Joseph Aspdin, a bricklayer from Leeds, experimented with a batch of concrete with an aggregate made from Portland stone, a form of limestone, which was quarried locally. This mixture was eventually granted a patent, patent number BP 5022, by the King of England, King George the fourth (Courland, 2011). After the death of Joseph, his son continued pioneering concrete into the 1840's earning him formal recognition as the creator of Portland cement. The 1840's also witnessed the invention and widespread use of reinforced concrete. Reinforced concrete, who's invention is credited to French gardener Joseph Monier, is composed of concrete inlaid with a stronger substance such as steel or iron which once cured provides increased tensile strength (Courland, 2011). Following the invention of Portland cement combined with the increase in tensile strength provided by iron or composite materials paved the road for an explosion of concrete production world wide.

What is Concrete and its Importance

As already determined, concrete is a composition of aggregates, water and hydraulic cement (Struble and Godfrey, 2004). Aggregate sizes determine, in part, the strength of the concrete; a coarse aggregate is responsible for weaker concrete, in contrast the finer the aggregate is the more bonds are formed and a stronger product is produced (Tsiskvili and Zabakhidze, 1970). Unfortunately, obtaining aggregate is not as simple as pulling rock and sand from the earth and there is a significant decrease of 'good' aggregates specifically in urban areas (Weizu, 2004). Fortunately, there is a possibility of using recycled materials as an alternative aggregate including old concrete, glass and rubber (Wang, 2004).

The most common hydraulic cement is portland cement or clinker, a tempered powder consisting of four minerals which react with water for strong bonds (Aïtcin, 2016). The four minerals that make this specific clinker blend are "tricalcium silicate (50 to 60%), the dicalcium silicate (20 to 25%, tricalcium illuminate (6 to 10%) and the tetracalcium ferroaluminate (6 to 10%)" and trace amounts of gypsum (Aïtcin, 2016). Other components may be used in conjunction with or in replacement of portland cement such as fly ash, slag, and silica fume. Fly ash is a by-product of burnt coal and can partially replace clinker up to 60% similarly another replacement of portland cement is observed with slag, a steel production by-product, replacing clinker up to 80% (Ahmaruzzaman, 2010; Yi et al., 2012). Silica fume, a by-product of silicon alloys, is similar to fly ash but is much finer; with finer particles there is a higher surface to volume ratio resulting in an ideal cement to facilitate stronger harder concrete (Siddique, 2011). Other components such as superplasticizers may be added into cements such as silica fume to reduce the amount of water needed while still maintaining a quality product (Mehta, 2004; Siddique, 2011).

Mixing proportions of these components can alter the structural proponents and performance of the concrete. For example, Malhotra (2002), compared mixing proportions of a HVFA concrete to determine the levels of strength produced in a hardened product. By altering various components of the concrete, while maintaining equivalent proportions of aggregate, Malhotra (2002), noted that between a low, moderate and high quality product, water appeared to be the most influential component as there was a mere 20 kg/m3 difference between a high and low strength concrete. Furthermore, both cement and fly ash had a ratio range of 100 kg/m3 to determine the strength quality of the concrete (Malhotra, 2002). Both moderate and high strength mixes require an additional superplasticizer, a water reducing additive, to maintain a low to cement ratio, to maintain a strong concrete with minimal water (Mahta, 2004). The flexibility of concrete construction allows for a plethora of uses from above ground construction to submersible applications.

Concrete is arguably the second most used component in the world, following water (Gagg, 2014). There are a variety of modern uses for concrete from architecture to transportation to culverts. Looking throughout history European countries such as Rome and Spain utilized concrete for buildings, bridges and dams during the Middle Ages (Gaudette, 2007). The use of concrete was not readily accepted as a building material in the United States of America until the nineteenth century but rather used for transportation (Gaudette, 2007). In 1818, the Erie Canal in New York was constructed of concrete and intended for transportation and in the 1920s the federal government became involved in the American's enthusiasm for good roads leading to many of today's U.S. routes (Gaudette, 2007). Another practical use for concrete comes from the way that it is utilized in aquatic settings, from drainage, cisterns and dams, concrete can be utilized to manipulate and guide the flow of water to allow for societal use or energy production. Concrete is not only used for its practical structure but also serves as an aesthetic style. Factory style utilitarian buildings boasted of exposed concrete inlay with glass during the early twentieth century, building interiors also followed the aesthetic design and embraced exposed concrete (Gaudette, 2007). Today, concrete has numerous possibilities from practical to aesthetic, encompassing more possibilities than those imagined by the inventor(s) themselves. With the ever expanding uses of concrete, engineers today consider not only how to improve concrete but how to forge concrete in a more environmentally friendly and sustainable way.

Environmental Impacts of Concrete

According to Ramsden (2020), concrete contributes to roughly eight percent of the total global CO_2 emissions. These emissions are not caused from the gathering of aggregates but rather from the processing which estimates that each pound of concrete releases 0.93 pounds of CO_2 (Ramsden, 2020). With an uprising in society's use of concrete, there is also a notable rise in increased temperatures in urban areas (Elizondo-Martinez et al., 2020). It is imperative to consider how to reduce or eliminate the CO2 emissions directly correlated with concrete if the globe is to continue its heavy use of the material.

Portland cement is the most widely used cement across the globe attributing 1.6 billion tons of cement to a total 12 billion tons of concrete and a release of one ton of carbon dioxide (CO2) into the atmosphere (Shah and Wang, 2004). There are three basic stages of concrete construction that exude harmful emissions: raw material preparation, clinker combustion and cement preparation (Barker et al., 2009). Considerable amounts of natural resources such as limestone and sand are needed to make cement, this initial extraction is energy intensive and uses a variety of fossil fuels (Shah and Wang, 2004). The process of firing mixed materials at 1450 °C (Ige et al, 2022) which decarbonizer limestone to calcium oxide through another high energy process is also a significant contributor to emissions (Ali et al., 2011). As a way to mitigate emissions, more durable construction materials are being used.

One aspect of emission reduction that can be considered is the viability of sustainable materials or practices. There are four stages to consider with estimating the impact concrete's life cycle has on the environment from production of material, construction, life cycle and demolition (Struble and Godfrey, 2004). It is well known that the extraction process is a leading cause of concrete's total emissions but the other stages of the process can also be reviewed and potentially modified to offset emissions. Construction uses some energy and creates waste and is being addressed with questions of sustainable development (Struble and Godfrey, 2004). The lifetime of concrete structures should also be considered as the durability of materials will directly impact future designs, repairs and renovations to maintain a required level of functionality (Struble and Godfrey, 2004). Finally, the demolition stage of concrete is where the most waste is created but with practices of sustainability, the reuse of old concrete as aggregates reduces the environmental impact (Struble and Godfrey, 2004). By reducing the need for replacement concrete structures by using more durable concrete, sustainable processes are reducing energy consumption and wastes production of the concrete lifecycle.

Overall, temperature fluctuations may be considered null when considering the vast impacts of climate changing CO2 being released into the atmosphere. Environmental issues are starting to rise as more of the world is paved, in 2002 it was estimated that three percent of the total Earth's surface was paved (Sinha et al., 2002). Natural processes such as the hydrological cycle causing runoff and water pollutions (Rodriguez-Hernandez et al., 2013) to rising urban temperatures are a direct result of increasing pavement, and their subsequent increased vehicle use impacts the environment in a more subtle way (Elizondo-Martinez et al., 2020). City temperatures are increasing due to pavement's solar absorption capabilities and creating urban heat islands (Elizondo-Martinez et al., 2020). In an attempt to mitigate and reduce these heat islands, concrete pavement aids by reflecting solar radiation; it also serves as a filter for aquifer recharge purposes (Elizondo-Martinez et al., 2020).

Alternative 'Concretes'

Faced with the scary reality of climate change the world will be forced to find alternatives that are more carbon friendly or even carbon neutral. This same reality can be seen in other aspects of society such as electrical generation with the introduction of wind/solar/hydro power and the increasing pressure to replace internal combustion engine vehicles with fully electric vehicles such as Teslas. Global concrete production results in eight percent of the total global CO2 emissions, with society's reliance on concrete only increasing the need for green alternatives has never been stronger. A sum total of eight alternatives are discussed in further detail throughout the book, however, a brief summary of each will provide the reader with an initial perspective into how each alternative may positively impact the environment.

Ferrock

Formed primarily of waste steel or iron dust/shavings and silica which can be sourced from crushed recycled glass, Ferrock is an environmentally friendly building compound that is nearly carbon neutral. Ferrock acts as a waste management tool as recycled materials are able to be efficiently used to produce a carbon neutral product. Raw materials of Ferrock include iron powder, metakaolin, limestone, oxalic acid and fly ash and can vary proportion depending on the strength level required (Das et al., 2015). Coming in at almost five times stronger than Portland cement, Ferrock is praised for its increased flexibility, making it a prime construction material in high seismic areas such as California (Gobinath et al., 2022).

Ashcrete

Combating the increasing amount of non-recyclable waste filling landfills or being shipped from western nations to less wealthy nations is becoming an ever increasing issue. This is where ashcrete as a technology fills the gap by being a product composed primarily of recycled material while extending its green impact by acting as a source of capturing carbon from the atmosphere. Fly and bottom ash are the primary ingredients in the production of ashcrete, both can be sourced from the burning of man-made waste such as household garbage. Waste to energy plants are used within the United States, the heat generated from burning the waste is used to create steam for the creation of electricity while the leftover ash is kept for use in ashcrete (Environmental protection Agency [EPA], 2022). Another advantage that ashcrete has over traditional portland concrete is the reduction in water required in the production of the concrete, this results due to the less reactive chemicals present in the ash.

Timbercrete

Similarly to ferrock and ashcrete, timbercrete derives its name from the root material used in the construction, which in this case is saw dust collected from large scale timber mills. Advantages include cheap supply costs as recycled sawdust is considered a waste product as well as its lightweight characteristics (Hammood, 2020). The increased versatility of timbercrete can also be related to the decreased emissions required in its transportation as it can be formed into any shape (Hammood, 2020). It is liked in the construction industry due to the building properties of wood while maintaining the strength and versatility of concrete. In addition to the environmental advantages of timbercrete it also acts as a carbon trap by preventing the cellulose in the saw dust from breaking down naturally which releases CO_2 or the alternative method of discarding waste saw dust which is burning (Timbercrete an Introduction, 2015).

Aircrete

Developed early in the 20th century with widespread use coming in the mid 20th century, Aircrete is produced through a mixture of various cementitious materials such as cement or pulverized fuel ash (PFA), lime, sand, water and aluminum oxide powder (Ahmed, 2017). Once cured in an autoclave for roughly ten hours, the final result is anywhere from sixty to eighty-five percent air. During the curing process, a chemical reaction occurs which results in the creation of Tobermorite which is why Aircrete has such a high compressive strength despite the abundance of large pores (Ahmed, 2017). An additional positive to the use of Aircrete is the increased thermal conductivity resulting from the increased air volume found in the final product. Making Aircrete a valuable buil;ding material for homes in colder climates (Callister, 2010).

Hempcrete

The early 20th century witnessed a boom in the industrial hemp business before a massive decline due to market competition. Before hemp lost its market influence it made an impact on the production of environmentally friendly concrete. Production requires drastically less energy and production of hemp is the definition of renewable (Jami et al., 2019). Lacking the structural strength of other alternatives available, Hempcrete is commonly used in flooring and the shells of buildings (Novakova & Sal., 2019). Hempcrete recovers from this lack of strength with its increased isolative properties resulting from the increased biomass found in the mixture (Barbhuiya & Bhusan Das, 2022).

Plastic Production

With nearly 6.5 billion tonnes of plastic waste and discarded rubber produced worldwide each year (Almeshal et al., 2020) reduction of such waste can happen in the form of plastic as concrete aggregate (Siddique et al., 2008). Plastics can be challenging to incorporate into concrete as they may contain harmful chemicals or have yet to undergo environmental risk assessments, however, the benefits of utilizing a challenging waste are immense (Siddique et al., 2008). Using plastics as aggregates would not only reduce overall waste in landfills but it has been determined that plastics allow for a more durable, cost effective and aesthetically pleasing concrete (Siddique et al., 2008). Another use of plastics, if a processed raw material is not used directly, is similar to fly ash in that a usable product is formed during the incineration of plastic (Siddique et al., 2008). Although plastic ash would not be the most environmentally friendly when considering emissions, it does solve the problem of excess physical waste as only a small fraction of plastics can be used as aggregate (Siddique et al., 2008).

Mycelium

As a relatively new organic material, mycelium has been recently explored as a valid option for developing sustainable concrete. Mycelium is understood as the main body of fungi or mushrooms but according to Shlyakhova and Yegorochkina (2022), mycelium can be made of wood, steel, aluminum or other concrete elements. Due to the high cost of mushroom farming mycelium is still in its early stages of development, and more specifically for the use of concrete, is still not considered as a preference to other environmentally friendly alternatives (Shlyakhova and Yegorochkina, 2022). Although cost is a potentially negative factor, mycelium is able to bond with concrete to form a self-healing concept (Luo et al., 2018). The theoretical application of such a concept would provide safe, natural, sustainable and pollution free repairs in concrete, further eradicating a significant challenge of concrete (Luo et al., 2018). Recent studies have chosen specific strains of mycelium that produce biomineralization with a faster forming spore and subsequently quicker self-healing quality with the intent to improve the durability of concrete (Zhang et al., 2021).

Bamboo

Considered to be a renewable and sustainable resource, bamboo is a versatile fast growing plant (Hebel, 2018). Bamboo itself is considered as a carbon sink, meaning it collects CO_2 while growing (Hebel, 2018). Due to its multipurpose nature, bamboo is an ideal material for sustainable material construction. Unlike some of the above listed materials, bamboo is used in concrete as an alternative reinforcement not a cement. Whole-culm (bars) and/or split bamboo make an ideal replacement for costly steel reinforcements (Archila et al., 2018). Bamboo has been used as a concrete reinforcement dating back a century to Asia, and further testing has commenced on the structural soundness of this organic material (Archila et al., 2018). In this early testing Glen (1950), noted that bamboo reinforcement was limited by its early brittle failure, reduced load capacity, bonding issues with moisture cracking and the need for asphalt additives. Modern research has proved that bamboo is an equivalent to steel reinforcement and has a tensile strength 1.5 times greater than its artificial counterpart (Kathiravan et al., 2021). To mitigate the remaining initial concerns of Glen (1950), modern civil engineers ensure to use dried bamboo which is coated with chemical treatments to reduce organic decays and allow for stronger bonds with cement compounds (Kathiravan et al., 2021). With modern testing and technologies applied, bamboo is a great alternative material for concrete reinforcement and serves to reduce concrete's carbon footprint.

Conclusion

The creation of concrete is one of the few inventions that have dramatically impacted the societies that utilized it, from the ancient Egyptians to the Romans to our modern era, concrete remains a constant. With the rapid population increase that society has undergone over the last two hundred years has brought with it an increased reliance on concrete that was once unimaginable. This has resulted in the production of concrete accounting for eight percent of the globe's total

CO2 emissions and a noticeable rise in temperatures in larger cities. The advancements of environmentally friendly options such as the ones explained in detail throughout the remainder of this book are just a select few examples of the exciting world of green concrete.

References

Ahmaruzzaman, M. (2010). A Review on the Utilisation of Fly Ash. Progress in Energy and Combustion Science, 36(3), 327-363. https://doi.org/10.1016/j.pecs.2009.11.003

Ahmed, A. (2017). Sustainable construction using autoclaved aerated concrete (aircrete) blocks. Research and Development in Material Science, 1(4). https://crimsonpublishers.com/rdms/pdf/RDMS.000518.pdf

Aïtcin, P. C. (2016). Portland Cement. In Pierre-Claude Aïtcin and Robert J Flatt (Eds.). Science and Technology of Concrete Admixtures (pp. 27-51). Woodhead Publishing. https://doi.org/10.1016/B978-0-08-100693-1.00003-5

Ali, M., Saidur, R., Hossain, M. (2011). A review on Emission Analysis in Cement Industries. Renew. Sustain. Energy Rev., 15(5), 2252-2261. https://doi.org/10.1016/j.rser.2011.02.014

Almeshal, I., Tayeh, B. A., Alyousef, R., Alabduljabbar, H., & Mohamed, A. M. (2020). Eco-friendly concrete containing recycled plastic as partial replacement for sand. Journal of Materials Research and Technology, 9(3), 4631–4643. https://doi.org/10.1016/j.jmrt.2020.02.090

Archila, H., Kaminski, S., Trujillo, D., Escamilla, E. Z., & Harries, K. A. (2018). Bamboo reinforced concrete: a critical review. Materials and Structures, 51, 101-119. https://doi.org/10.1617/s11527-018-1228-6

Barbhuiya, S., & Bhusan Das, B. (2022). A comprehensive review on the use of hemp in concrete. Construction and Building Materials, 341, 127857. https://doi.org/10.1016/j.conbuildmat.2022.127857

Barker, D. J., Turner, S. A., Napier-Moore, P. A., Clark, M., & Davison, J. E. (2009). CO2 Capture in the Cement Industry. Energy Procedia, 1, 87-94. https://doi.org/10.1016/j.egypro.2009.01.014

Brewer, J. (n.d.) The History of Concrete. Retrieved January 21, 2023, from http://matse1.matse.illinois.edu/concrete/hist.html

Callister, W. (2010). Materials science and engineering-An Introduction (5th edn.). https://ia801305.us.archive.org/29/items/MaterialsScienceAndEngineering/Materials%20Science%20and%20Engineering%20-%20Callister.pdf

Courland, R. (2011). Concrete planet : the strange and fascinating story of the world's most common man-made material. Amherst, N.Y.: Prometheus Books.

Das, S., Hendrix, A., Stone, D., & Neithalath, N. (2015). Flexural fracture response of a novel iron carbonate matrix - Glass fibre composite and its comparison to Portland cement-based composites. Construction and Building Materials, 93, 360-370. https://doi.org/10.1016/j.conbuildmat.2015.06.011

Diamond, J. M. (2011). Collapse: How societies choose to fail or succeed. Penguin. https://www.worldcat.org/title/collapse-how-societies-choose-to-fail-or-succeed/oclc/919700581

Elizondo-Martinez, E., Andres-Valeri, V., Jato-Espino, D., & Rodriguez-Hernandez, J. (2020). Review of porous concrete as multifunctional and sustainable pavement. Journal of Building Engineering, 27(1), 1-9. https://doi.org/10.1016/j.jobe.2019.100967

Environmental Protection Agency. (2022, March 16). Energy Recovery from the Combustion of Municipal Solid Waste (MSW). https://www.epa.gov/smm/energy-recovery-combustion-municipal-solid-waste-msw#:~:text=The%20ash%20that%20remains%20from,facilities%20recover%20energy%20from%20landfills.

Gagg, C. (2014). Cement and concrete as an engineering material: An historic appraisal and case study analysis. Engineering Failure Analysis, 40, 114–140. doi:10.1016/j.engfailanal.2014.02.00

Gaudette, P. E. (2007). Preservation of Historic Concrete: Problems and General Approaches, Revised. Government Printing Office. https://books.google.ca/books?hl=en&lr=&id=oY6Fc4o-BcMC&oi=fnd&pg=PA3&dq=historic+concrete+structures&ots=-2c1_TJcYx&sig=Kx1wNzh0L9q5jXRQ6f1g1kQFo0U&redir_esc=y#v=onepage&q=historic%20 concrete%20structure &f=false

Glen, H. E. (1950). Bamboo Reinforcement in Portland Cement Concrete. Clemson.

Gobinath S., Ramesh R. L., & Prajwal S. Desai, (2022). Performance of composite building materials using granite slurry, earth blocks, and ferrock. AIP Publishing, 2615(1), n.p. https://doi.org/10.1063/5.0117038

Hammood, Z. A. (2020). Using Sustainable Materials to Develop the Buildings to be Green. International Journal of Advanced Science and Technology, 29(3). http://sersc.org/journals/index.php/IJAST/article/view/31555

Harrison, T., Jones, M., & Lawrence D. (2019). The Production of Low Energy Cements. In P. Hewlett & M. Liska (Eds.), Lea's Chemistry of Cement and Concrete (Fifth Edition) (pp. 341-361). Butterworth-Heinemann. https://doi.org/10.1016

Hebel, D. (2018, January 31). Natural building materials: Bamboo. RICS. https://www.rics.org/en-in/news-insight/future-of-surveying/sustainability/natural-building-materials-bamboo/

Ige, O. E., Olanrewaju, O. A., Duffy, K. J., & Collins, O. C. (2022). Environmental Impact Analysis of Portland Cement (CEM1) Using the Midpoint Method. Energies, 15, 1-16. https://doi.org/10.3390/en15072708

Jahren, P., & Suit. (2017). History of concrete: A very old and modern material. World Scientific Publishing. https://doi.org/10.1142/10172

Jami, T., Karade, S. R., & Singh, L. P. (2019). A review of the properties of hemp concrete for green building applications. Journal of Cleaner Production, 239. 10.1016/j.jclepro.2019.117852

Kathiravan, N. S., Manojkumar, R., Jayakumar, P., Kumaraguru J., & Jayanthi, V. (2021). State of art review on bamboo reinforced concrete. Materials today: proceedings, 45(2), 1063-1066. https://doi.org/10.1016/j.matpr.2020.03.159

Luo, J., Chen, X., Crump, J., Zhuo, H., Davies, D. G., Zhuo, G., Zhang, N., & Jin, C. (2018). Interactions of fungi with concrete: Significant importance for bio-based self-healing concrete. Construction and Building Materials, 164(10), 275-285. https://doi.org/10.1016/j.conbuildmat.2017.12.233

Malhotra, V.M. (2022). High-Performance, High-Volume Fly Ash Concrete. Concrete International, 24(7), pp. 30-34.

Mehta, P. K. (2004). High-Performance, High-Volume Fly Ash Concrete for Sustainable Development. In Kejin Wang (Eds.). Proceedings of the International workshop on sustainable development and concrete technology (pp. 3-14). Iowa State University. https://publications.iowa.gov/2941/1/SustainableConcreteWorkshop.pdf#page=14

Novakova, P., & Sal, J. (2019, September). Use of technical hemp for concrete-Hempcrete. In IOP Conference Series: Materials Science and Engineering (Vol. 603, No. 5, p. 052095). IOP Publishing. https://iopscience.iop.org/article/10.1088/1757-899X/603/5/052095/pdf

Ramsden, K. (2020, November 3). Cement and Concrete: The Environmental Impact. Princeton Student Climate Initiative. https://psci.princeton.edu/tips/2020/11/3/cement-and-concrete-the-environmental-impact

Rodriguez-Hernandez, J., Fernandez-Barrera, A. H., Andres-Valeri, V. C. A., & Vega-Zamanilla, A. (2013). Relationship between Urban Runoff Pollutant and Catchment Characteristics. Journal of Irrigation and Drainage Engineering, 139(10), 833-840. https://doi.org/10.1061/(ASCE)IR.1943-4774.0000617

Robertson, D. (1969). Greek and Roman Architecture. Cambridge University Press. https://books.google.com.gi/

Chapter 2: Ferrock
by Zarish Jawad

Climate change makes our green and beautiful world become greyer every year. Besides so many other factors, the construction industry is one of the significant contributors to climate change. It is because the construction industry is full of high-generation carbon materials and processes that continuously contribute to climate change (Bonnefin, n.d.). It is estimated that concrete or cement is the most commonly consumed substance on Earth after water. Worldwide, 5% of man-made emissions come from the production of cement. Moreover, concrete, regardless of its durability and versatile qualities, is a major contributor to the emission of greenhouse gasses, thus making up 8% of overall global emissions (Denus, 2021). Currently, around 4 billion tons of concrete are produced yearly, created from usually unsustainable raw materials that release carbon dioxide into the air (Bonnefin, n.d.).

Though much attention has been paid and diverted over the last years towards factory farming, air transport, and other forms of carbon emission, more attention should be paid to the construction industry. However, governments, citizens and, above all, researchers have become more mindful and aware of the impact of rapid construction on the environment (Denus, 2021). Due to this, the construction industry is going through big changes in terms of more eco-friendly building materials. Against the same backdrop, many new building materials and innovative technologies have emerged intending to improve current environmental conditions, including materials such as Ashcrete, Timbercrete, and Ferrock (Kerr et al., 2022). Amongst the alternatives, Ferrock aims to replace or reduce the use of concrete/cement with a carbon-negative and better-performing material composed primarily of recycled materials.

What is Ferrock?

Ferrock is a carbon-negative, environmentally friendly compound and building material used as an alternative to and a substitute for concrete and cement. It is cheap to manufacture and comprises 95% recycled materials such as scrap iron and silica from crushed glass (Kerr et al., 2022). Ferrock consists of three main components: iron-rich ferrous rock, waste steel dust and silica from round-up glass. When mixed, the ingredients undergo a chemical reaction resulting in a strong, hard solid form like concrete. Steel dust reacts with carbon dioxide to form iron carbonate, which becomes Ferrock after solidification. The hardening process occurs when a mixture of steel dust and silica is mixed with iron rock and water and exposed to high carbon dioxide concentrations (Liu et al., 2022). It is not only five times stronger than concrete or cement. Still, it is also more flexible, allowing it to be used in highly active environments with a concern for seismic activity and it can endure large compressive stresses caused by seismic forces.

History & Invention

Ferrock was invented as an alternative to Portland cement by Dr. David Stone, founder and the owner of Iron Shell Media Technologies, in 2002. He was a former Phd student at the University of Arizona in the Soil, Water and Environmental Science Department. Stone, while researching ways to prevent iron from rusting and hardening, accidentally made Ferrock (Liu et al., 2022). Initially, Stone abandoned the test and experiment but later changed his mind and focused on pursuing and finding a material with similar physical capabilities as concrete but more eco-friendly. After figuring out how to manufacture ferrock and test his idea, he worked with the Tohono O'odham Nation Reservation to source the silica he required to undertake his tests (Kerr et al., 2022). He also received $200,000 worth of Environmental Protection Agency (EPA) grants for his research and work, which permitted and facilitated him to create demonstrative projects. In 2013 the US Patent and Trademark Office issued a patent for the University of Arizona for Ferrock. Stone 2014 worked out a contract to hold a license in collaboration with Tech Launch Arizona (TLA) to commercialize his invention (Liu et al., 2022). Ferrock is now a trademarked name, and currently, the University of Arizona holds a patent for Ferrock. An exclusive license has been given to Iron Shell Material Technologies, a company founded by Dr. Stone that allows for the commercialization of Ferrock but also its manufacturing and selling (Vijayan et al., 2020).

How is Ferrock Made?

The basic building blocks of Ferrock come from iron discards in the form of waste steel dust as the iron in waste steel dust reacts with carbon dioxide to form iron carbonate. The iron carbonate becomes part of the material's mineral matrix, adding to its overall strength. By using the steel dust byproducts which come from multiple types of industrial processes, Ferrock is formed (Kerr et al., 2022). Generally, this dust is thrown away, thus wasting its potential as this leads to heavy metals in landfills, thus causing toxins to seep into the ground. Apart from the steel dust waste, silica is also needed (made from ground-up glass). To produce Ferrock, steel dust and silica are mixed with ferrous rock (an iron-rich mineral) and other ingredients which cause rusting or corrosion (Liu et al., 2022). After which water is mixed in, this mixture turns into a paste similar to the chalky glop of cement. Ferrock can be used exactly like cement or concrete when mixed up. It can be poured into any mold or trowel with it. Lastly, the mix gets exposed to carbon dioxide. The iron in the dust will start to rust as it absorbs carbon dioxide. The CO2 gets amalgamated into the mixture, thus forming iron carbonate. In a week's time, the material hardens into solid form known as Ferrock. It traps the carbon dioxide into the material as the Ferrock gets hardened. The addition of carbon dioxide makes the Ferrock stronger (Bonnefin, n.d.).

Ferrock V/S Concrete

One of the most commonly used types of cement in concrete production is Portland cement. However, when compared with Portland cement, Ferrock is five times stronger and far more flexible. Ferrock can also handle more compression as compared to concrete material thus making it more resistant to breakages and cracks (Kerr et al., 2022). Ferrock is also far more sustainable than concrete as the main component of concrete which is Portland cement has a huge carbon footprint as its production is responsible for 8% of the world's carbon dioxide emissions. On the other hand, Ferrock eradicates the need for any cement and relies on the waste steel dust. Besides, it absorbs carbon dioxide from the air instead of adding more to the environment. A best quality of Ferrock is that exposure to saltwater environments acts as a strengthening agent for Ferrock, which makes it an ideal building material for construction projects in marshes and other coastal construction sites that are heavily exposed to saltwater. Given the similar functional properties, Ferrock is a promising alternative to cement (Kerr et al., 2022). The cement curing process needs approx 7 to 28 days of hydration and concrete takes 24 to 48 hours to harden, but it takes up to 28 days to reach its full strength. On the other hand, Ferrock cures within four days of carbonation and requires at least a week to expand.

Benefits & Advantages of Ferrock

There are many advantages of using Ferrock for construction purposes instead of cement or concrete, such as its strength, carbon neutrality, flexibility, chemically inactive, environmental friendliness and durability (Vijayan et al., 2020).

Ferrock is Strong

Ferrock hardens into a solid form, making it like concrete, but once hardened, Ferrock is five times stronger than concrete. Thus it can resist more weight, compression and damage without being destroyed. The strength of Ferrock is partly due to the presence of iron carbonate, a product of the chemical reaction between the steel dust waste's iron and the CO_2. Ferrock has strengths ranging from 5,000 to 7,000 psi, while some Ferrock tests reached 10,000 psi (Vijayan et al., 2020).

Flexibility

Besides being strong, Ferrock has some flexibility, which allows Ferrock to withstand more movement and pressure without cracking. In comparison, concrete is completely solid, any minor or slight movement can cause cracks that can weaken the complete structure. Withstanding some movement without damage makes Ferrock great for use in seismic areas, like earthquakes or compression. Moreover, Ferrock is four times stronger during flexural tests than Portland cement (Vijayan et al., 2020.

Environmentally Friendly

Ferrock has recycled materials as it is created from steel dust waste and silica(95% of Ferrock consists of recycled material). Steel dust waste is a byproduct of many industrial processes, while silica comes from the ground-up glass (Vijayan et al., 2020). In addition, Ferrock is also considered carbon neutral, which means it does not emit much carbon dioxide during manufacturing(though it does put out a bit of carbon dioxide during production, like cement and concrete, however lower than the amounts emitted by either). Producing Ferrock can reduce greenhouse gasses. In liquid form, Ferrock uses CO_2 to help it harden (Liu et al., 2022). When the mixture of steel dust waste, silica, Ferrock rock, and water is exposed to air, the iron in the steel dust waste absorbs and reacts with carbon dioxide from the air. Carbon dioxide mixes into the mixture, thus trapping the gas inside the rock as it turns into a solid. Therefore, Ferrock works as a carbon dioxide filter, removing some of the carbon dioxide in the atmosphere. Ferrock uses absorbed carbon dioxide to form its final shape, a sheet of solid hard Ferrock (Kerr et al., 2022).

Ferrock is Durable

Ferrock is considered relatively chemically inactive. When exposed to gasses or chemicals, Ferrock would not deteriorate or degrade. On the other hand, concrete can get worse over time and with exposure to chemicals. In contrast, Ferrock is resistant to rotting, oxidation, UV radiation, corrosion and rust. Nor can it be damaged by chemicals, thus making it a viable option and an ideal material for tubes and pipes (M, N. et al., 2021). In addition, Ferrock is even strengthened by exposure to saltwater as it is resistant to the effects of saltwater. Therefore, it is an excellent choice of material for underground environments and marine construction projects.

DrawBacks of Using Ferrock

Given the benefits and advantages listed above, Ferrock is the alternative. It can be used in place of concrete or cement, especially because it can mitigate environmental issues and climate change. However, there are some disadvantages of using ferrock, which make it unsuitable for extended use or in major construction projects. Below are a few such disadvantages.

Untested & New Material

What prevents Ferrock from replacing concrete or cement entirely is that it is untested, still there is plenty of information that has not been figured out. Therefore, Ferrock is a relatively new material as it has only been around since the early 2000s. It has yet to be seen and used much in the industrial field (M, N. et al., 2021). While in contrast to Ferrock, concrete has been around for over 200 years, tested and tried by most developers and builders. Though we know Ferrock is strong enough to resist more weight and compression than concrete or cement. Thus far, the overall capabilities of Ferrock remain unproven. It is because we do not know how Ferrock behaves under a wide range of building conditions as it needs specific techniques to harden or how long the material's life cycle is as it is also not clear that we can use the same types of concrete methods on Ferrock.

High Cost

Another factor which still puts concrete or cement ahead in most construction projects are the cost. Though, as of now Ferrock is not being widely used and it is still very easy to find the material required to create Ferrock for a low cost. However, suppose companies realize that they can make a profit through their waste as waste is required to manufacture Ferrock (Vijayan et al., 2020). In that case, it could increase costs as it will make finding the necessary waste materials more challenging. Therefore despite the growing environmental issues associated with the greenhouse gases caused by concrete manufacturing, over time, it could become too expensive to use Ferrock over concrete or cement. Ferrock is not a cost effective solution for large-scale projects such as highways and road development (Liu et al., 2022). However, Ferrock is more suitable for niche products. Suppose the steel dust goes directly from being a waste to being a useful building material. In that case, the cost of producing Ferrock will be exponentially high, which makes the construction process all the more costlier. In Ashwin Reddy, Director, Aparna Construction words, "Ferrock is not a cost-effective construction material in the long-term; however, it is a starting point to address the insatiable hunger for development and the devastating results it has on the environment."

Materials

We already know that 95% of the ingredients of Ferrock are recycled material, which means that natural resources are not utilized to make ferrock, unlike concrete. Ferrock is an eco-friendly alternative to using cements or concrete. Production of Ferrock mainly depends on the creation of other goods, like silica, steel dust waste and metal shaving (Liu et al., 2022). The two main ingredients in Ferrock are waste steel dust and silica, which are the byproducts of other construction processes. If Ferrock were to become a mainstream construction material then the sufficient availability of the ingredients, steel dust, silica and metal shaving is must. Out of the three ingredients, steel dust and silica are the byproducts of other construction processes. Moreover, it also takes a lot of silica and metal shavings to create Ferrock, and with both of those in limited supply, it makes Ferrock unsuitable for big projects (M, N. et al., 2021). It means that manufacturers of Ferrock based on the availability of ingredients are limited in how much rock they can make. Due to the limited supply of materials and ingredients required to manufacture Ferrock, it is not economical to use in especially for large projects such as roads and highways.

Ferrock-Application & Uses

As people are becoming more aware of climate change, environmental issues and above all the growing popularity of Ferrock manufacturing, the use of Ferrock has become an attractive option. It is because the

Ferrock material has many uses for sustainability and represents a great alternative (M, N. et al., 2021). Currently Ferrock has been approved and tested to be used for larger and smaller items and projects like in slabs, bricks, pavers, paving, breakwaters, sidewalks, benches and walls. However there have not been significant advancements for Ferrock approval and use for large structures and projects like construction of roads. However there have been some promising results and opening for Ferrock as far as marine based environment and construction is concerned. In view of its amazing ability to get more durable with exposure to sea and salt water, Ferrock is chemically relatively inert, making it suitable for offshore projects (M, N. et al., 2021). Moreover sea salt increases the strength of ferrock. Therefore Ferrock can be used in marine environments for seawalls, piers, pilings, foundations and breakwaters. Ferrock, resistant to chemicals found in sewage like sulfuric acid, is an excellent material for constructing pipes and tubes as it is resistant to oxidation, UV light, corrosion, chemicals, rot and rust. At present as there is limited usage of Ferrock in the construction industry, you will find and come across very few builders having the experience of using Ferrock in their projects (Liu et al., 2022). However, with more awareness and availability of Ferrock in the market, its usage will definitely increase in the future.

Will Ferrock Replace Concrete/Cement Soon?

Ferrock can bear more compression before breaking and is far more flexible than concrete or cement, thereby resisting earth movements caused by seismic activity or industrial processes. But still at present or even in the near future, the chances of Ferrock completely replacing concrete/cement are slim (M, N. et al., 2021). It's because concrete is not only relatively cheap to purchase, but it is also easy to produce, besides it can be used to build a wide variety of structures. Concrete can also be proven effective when it comes to neutralizing carbon dioxide emissions (Vijayan et al., 2020). For example magnesium silicate-based cement can also not only absorb large amounts of carbon dioxide as it hardens but the abundance of magnesium silicates also makes it a lower-cost alternative to Ferrock. Having mentioned that, one should consider the usefulness of Ferrock. Instead, Ferrock is an eco-conscious alternative because when combined with other eco-friendly materials and building strategies, it becomes possible to lower overall greenhouse gas output and create more sustainable environmental conditions (M, N. et al., 2021). Therefore, it is more beneficial to replace concrete and cement with Ferrock.

Ferrock-Present & Future

Though Ferrock has been used in smaller construction sites, it still has a long way before Ferrock is widely adopted. To test and demonstrate the capability and strength of the Ferrock in commercial projects many field experiments have been conducted (Vijayan et al., 2020). Ferrock will likely grow within marine environments before it is widely accepted and adopted. The large-scale adoption of alternative materials to cement or concrete like Ferrock remains an uphill battle, especially when the cement and concrete that have been in use for over 200 years are time tested (Kerr et al., 2022). In contrast, unfortunately, there is not enough data to prove the same about alternative materials, particularly Ferrock, no matter how sustainable they are and their ability to survive in the long term. However, in the future more methods and procedures will be explored by researchers to help reduce greenhouse gas emissions and as they are finding carbon-neutral building materials, the demand and utility of Ferrock may grow.

Conclusion

Considering the rising cost of cement and its adverse environmental effects, alternative construction materials such as Ferrock seem to be a viable option and solution. It is a carbon-negative, recycled and durable material that could replace cement or concrete in the future. Given the changes in global warming over the last few hundred years, it is high time to find ways to reduce our carbon footprint. The only suitable way to do that is to produce and use eco-friendly materials made of recycled materials. Therefore, realtors must adopt more cost-effective and nature-friendly building materials such as Ferrock. Made from recycled materials, Ferrock can absorb carbon dioxide out of the atmosphere, reducing greenhouse gasses. Furthermore, Ferrock is a strong, flexible, green construction material that can be used as an alternative to cement and concrete. In addition, there is a dire need to continue researching and exploring new building materials that are more sustainable than those in use at present and greatly contribute to global greenhouse emissions.

References

Bonnefin, I. (n.d.). Emerging materials: Ferrock. Certified Energy. Retrieved January 27, 2023, from https://www.certifiedenergy.com.au/emerging-materials/emerging-materials-ferrock

Denus, M. (2021, June 22). All about ferrock. Medium. Retrieved January 27, 2023, from https://amastgroup.medium.com/all-about-ferrock-c390b27192d1

Kerr, J., Rayburg, S., Neave, M., & Rodwell, J. (2022). Comparative analysis of the global warming potential (GWP) of structural stone, concrete and steel construction materials. Sustainability, 14(15), 9019. https://doi.org/10.3390/su14159019

Liu, T., Chen, L., Yang, M., Sandanayake, M., Miao, P., Shi, Y., & Yap, P.-S. (2022). Sustainability considerations of Green Buildings: A detailed overview on current advancements and future considerations. Sustainability, 14(21), 14393. https://doi.org/10.3390/su142114393

M, N. ., Manjunath, Y. M. ., & Prasanna, S. H. S. . (2021). Ferrock: A Carbon Negative Sustainable Concrete. International Journal of Sustainable Construction Engineering and Technology, 11(4), 90–98. Retrieved from https://publisher.uthm.edu.my/ojs/index.php/IJSCET/article/view/8084

Vijayan, D. S., Dineshkumar, Arvindan, S., & Shreelakshmi Janarthanan, T. (2020). Evaluation of ferrock: A greener substitute to cement. Materials Today: Proceedings, 22, 781–787. https://doi.org/10.1016/j.matpr.2019.10.147

Chapter 3: Ashcrete
by Sudipta Samadder

What is Ashcrete?

Ashcrete is an environmentally-friendly construction material that is considered a viable substitute for concrete. An estimated 93% of all the materials used in the production of Ashcrete are recycled material. With an average density of 1.8092 g/cm3, hardened Ashcrete has a low permeability and a compressive strength exceeding 21 MPa (Beyondhomes, 2020). Ashcrete also requires a very low energy-intensive manufacturing process compared to traditional concrete. Additionally, it is not a widely used substance but it is used the most in the United States. The product is also known for having roughly twice the strength of Portland cement (Beyondhomes, 2020). It has been shown that Ashcrete can in fact trap carbon dioxide from the air, reducing carbon emissions (What Is Ashcrete? - The Constructor, n.d.). Moreover, Ashcrete has higher compressive and tensile strength than concrete, and it is more acid- and fire-resistant (Material District, 2019). Traditional concrete consists of cement, sand, gravel, and water, with cement being the main drawback that confers a negative environmental impact. In comparison, Ashcrete is made of a mixture of fly ash, borate, bottom ash, and chlorine compound, with Class C fly ash being the main component (What Is Ashcrete? - The Constructor, n.d.). Hence, Ashcrete is often referred to as high-volume fly ash concrete.

History of Ashcrete

Ashcrete was first discovered by the Romans as they were the first to experiment with an early form of acetate by using volcanic ash to build aqueducts and historic structures. More recently, Ashcrete was created by Pliny Fisk III as a means of mitigating both the high amount of carbon dioxide produced during cement production and the disposal of fly ash, a residue of coal-based energy production (Nair, 2021). It all started in his kitchen which he would call his 'earth lab'. Fisk was performing a teacup experiment where he was scooping a couple of spoonfuls of water into a teacup filled with fly ash from a coal-fired power plant. In nearly twenty-minutes, Fisk realized that the fly ash had hardened and turned into something that could not be broken with hands (Lerner, n.d). He decided to test this further by making a proper mix which resulted in a substance that became so hard that it broke the compression tester. Later, it was found that this strong and hard material was tested out at 6000 psi which is about twice the strength of Portland cement. Fisk soon came up with a recipe for this alternative cement material which he named AshCrete (Lerner, n.d). He would present his new findings at various offices and pass out samples to officials at the Department of Energy (DOE), explaining their benefits over concrete.

Fly Ash

Fly ash is not a new technology in construction, but it is one that is growing rapidly! Prior to air pollution control standards, fly ash was generally released into the atmosphere, but now it must be captured by pollution control equipment. There are two classes of fly ash defined in the Standard Specification for Coal Fly Ash and Raw or Calcined Natural Pozzolan for Use in Concrete (ASTMC618). These include Class F fly ash and Class C fly ash. If the ash produced by burning anthracite or bituminous coal meets the chemical composition and physical requirements specified in ASTM C618, it is referred to as class F fly ash (Coal Fly Ash, 2016). This type of ash is typically pozzolanic, meaning they contain glassy silica and alumina that will, in the presence of water and free lime, react with the calcium in the lime to produce calcium silicate hydrates (cementitious compounds) (Coal Fly Ash, 2016). If the ash produced from burning lignite or subbituminous coal has pozzolanic properties as well as self-cementing properties (ability to harden and gain strength in the presence of water alone) in addition to meeting the chemical and physical requirements outlined in ASTM C618, it is referred to as Class C fly ash (Coal Fly Ash, 2016).

There are several advantages to using fly ash in Ashcrete. Pozzolanic compounds in both types of fly ash add to the strength, impermeability, and sulfate resistance of Ashcrete, and also reduces the maximum rise

in temperature (What Does a Pozzolan Do in the Concrete?, n.d.). Moreover, pozzolanic ash decreases water demand as the cementitious material fills the spaces between cement grains and displaces trapped water. Another advantage of pozzolans is their low cost as they are simply a by-product of coal combustion. Class F fly ash particles are typically covered in a kind of melted glass which significantly reduces the risk of expansion due to sulfate attack in fertilized soils or near coastal areas (Gomaa et al., 2017). Class C fly ash is also resistant to expansion from chemical attacks. Compared to Class C, Class F fly ash has little to no cementing value because it is generally low-calcium and has a carbon content that is usually less than 5 percent. Class C fly ash serves as a more suitable ingredient in the production of Ashcrete as it is typically composed of high-calcium fly ashes with a carbon content of less than 2 percent. If Class F fly ash is used, it is used at dosages of 15 to 25 percent by mass of cementitious material whereas Class C fly ash is used at dosages of 15 to 40 percent. Compared to traditional Ashcrete, Class C fly ash helps to reduce cracking and bleeding of fresh Ashcrete due to its low heat of hydration (Facilitator, 2022). This results in lower permeability and greater durability of Ashcrete. Additionally, it increases the resistance of Ashcrete against temperature fluctuations and corrosion! Other advantages of fly ash include resistance to acid and sulphate attacks, conservation of natural resources from reduction in the requirement of clay, sand, and limestone in Ashcrete, and lowers cement requirement. These advantages make fly ash suitable for multiple applications. As a pozzolanic material, fly ash is commonly used by the cement industry for making Portland Pozzolana Cement since its SiO_2 and $Al2SiO3$ contents closely match Portland cement (Facilitator, 2022). Fly ash is also used in road and embankment construction. It saves topsoil which avoids the creation of low-lying areas. Bricks can also be made from material that can provide a certain amount of plasticity, and fly ash is one of them. The use of fly ash with soil helps reduce the cost of building material (Facilitator, 2022)
.

Apart from the advantages, there are a few disadvantages of using fly ash. One is the slower strength and longer setting time of Ashcrete. Due to the presence of fly ash, Ashcrete takes longer than cement concrete to reach maximum strength, so construction time may be extended. A second disadvantage is seasonal limitations (Nair, 2021). Ashcrete is more susceptible to low temperatures, which significantly lengthen setting times and strength gain. Many regions even ban the use of fly ash in winter. Furthermore, the amount of air-entraining required for fly ash concrete is higher. A third disadvantage is sustainability. Fly ash is a surplus waste product made from coal production. Although it reduces global warming and pollution, it still relies on unsustainable methods of production in the long run (Nair, 2021). As the use of fly ash in construction becomes more widely used, it may lead to an increase in demand for combustion of coal. Furthermore, currently Ashcrete contains a hazardous chlorine compound. The search for an alternative to this chlorine-based compound is ongoing. The issue of fly ash storage is also in the process of being reviewed (Nair, 2021).

Borate

Another ingredient that is used in the production of Ashcrete is borate. Borates are naturally-occuring minerals containing boron. Recognized as the fifth element in the periodic table, boron does not exist by itself in nature but rather combines with oxygen and other elements to form boric acid, or inorganic salts called borate. Rock, soil and water contain trace amounts of boron. The element is required for plant growth and is also an important part of a healthy diet for humans (Material Insights, n.d.). It has a crucial function in industrial construction as it is a strengthening additive in concrete, steel, and aluminum alloys for constructional support.

There are several advantages to using borate in construction. Some of its applications are observed in cement. Several types of cement are produced with the help of borate, especially those with a high amount of belite. This is because borates promote the formation of dicalcium silicate. As a result of the use of belite cement, compressive strength of concrete and its alternatives also have been improved (Moravej et al, 2022). One significant advantage of using borate is that it extends the period of cement hydration. Due to the fact that borates have the ability to function as a set retardant, construction workers are able to better control the pouring of cement and direct it to its designated location - before premature setting occurs. As a result, boron compounds are used in the production of Ashcrete as effective set-retarding additive materials, similar to hydroxylated carboxylic acid, lignins, sugar, and some phosphate compounds. For applications such as capping orphaned oil wells, where cement must reach deep pockets before setting, this capability of greater control is especially useful. In infrastructure, more than $4 billion is allocated for well capping (Moravej, 2022). Ceramic tile bodies now incorporate borate as an essential ingredient, allowing manufacturers to increase productivity and decrease energy consumption (Material Insights, n.d.). Another advantage of using borate is that they make the production process more efficient. Through its glass-forming and fluxing agent properties, they lower the temperature from 1500 degrees celsius to 1300 degrees celsius and improve melt viscosity (Moravej, 2022). As conventional retarders and auxiliary admixtures for high-temperature viscosity, borate compounds have been commonly used in oil-well cementing operations (Davraz, 2010).

Bottom Ash

Coal bottom ash is another component in Ashcrete that is formed in coal furnaces. Ash from the bottom of the furnace is a coarse, granular, incombustible by-product of coal combustion. Most bottom ash is produced at coal-fired power plants, which are facilities that burn coal to make steam in order to generate electricity (Energy Education, n.d.). During the burning of pulverized coal in a bottom boiler, most of the unburned material of the coal is caught in the flue gases, and thus, is captured as fly ash. Only about 10-20% of this ash is bottom ash while the remaining percent is fly ash. The bottom ash is about the size of sand and is dark grey in colour. In the bottom of the furnace, there is a water-filled hopper that collects the ash and uses high pressure water jets to remove it from the hopper. The waste is then collected in a pond, and afterwards, it is disposed of or used after it is recycled by being stored in a storage area (Energy Education, n.d.). Bottom ash is mainly composed of silica, alumina, and iron with small traces of calcium, magnesium, and sulfate.

Coal bottom ash is often used as a replacement for natural sand in concrete and its alternatives. This can create a variety of effects on multiple properties of concrete and alternatives including their microstructure, weight, density, and compressive strength. In a study by Yüksel and Genç, it was found that replacing natural sand with bottom ash in concrete changes its network structure (Yüksel and Genç, 2007). By using bottom ash instead of natural river sand, which consists of irregular grains, the grains become circular and the pores become smaller and more widely distributed. Detachment of grains from the network structure increases as the amount of sand replaced by bottom ash increases (Yüksel and Genç, 2007). It is these discrete grains that make up the microstructure of bottom ash. Additionally, various research shows that there is an apparent decrease in unit weight of concrete or alternatives of concrete when natural sand is substituted with bottom ash. Compared to natural sand, bottom ash weighs less and has a porous structure, resulting in a decrease in the unit weight of concrete. Another factor that contributes to the lower density of bottom ash concrete has to do with the requirement for more water during mixing, which leads to a greater number of pores, larger pores, and thus the porous structure of the concrete (Singh and Siddique, 2013). In a study by Andrade et al, it was found that the use of bottom ash, with a specific gravity of 1.67 g/cm3 and a fineness modulus of 1.55, as a replacement for sand in concrete, resulted in a 25% decrease in densities of concrete, from 2170 kg/m3 to 1625 kg/m3 (Andrade et al, 2009). In an alternative study by Kim and Lee, the effect of fine and coarse bottom ash on density of concrete was studied (Kim and Lee, 2011). According to their studies, hardened concrete densities decreased linearly as the proportion of fine versus coarse bottom ash was increased. Bottom ash also affects the compressive strength of concrete and its alternatives. In a study by Chun et al, the researchers discovered that the strength of concrete differed based on the amount of pond ash (a mixture of dry fly ash and bottom ash) collected from each disposal site (Chun et al, 2008). Compared to normal concrete, there has been a relatively greater increase in compressive strength with a greater amount of pond-ash content. Such a trend might be a consequence of a decrease in the water/cement ratio resulting from the absorption of water.

There are some disadvantages to the incorporation of bottom ash. Based on where coal was mined, coal ash (including bottom ash) may have a different chemical makeup. One issue with bottom ash is that it exists in massive quantities. The presence of acidic, toxic, and radioactive matter in coal ash makes it a pollutant to be considered. A significant amount of bottom ash and other components of coal ash increases a person's risk of developing cancer and other respiratory diseases, according to the Environmental Protection Agency (EPA). This ash can contain lead, arsenic, mercury, cadmium, and uranium (Energy Education, n.d.). Additionally, the most common area where bottom ash is often found is in storage lagoons. Often, groundwater contamination occurs when fly ash and bottom ash are disposed of in lagoons without proper liners to prevent leaks and leaching. If the arsenic contaminated water is ingested by a person, it can put him or her at risk of developing cancer. The inhalation of bottom ash is less of an issue than that of fly ash, since bottom ash is much heavier than fly ash. Ingestion of bottom ash, however, can have adverse effects on the nervous system, cognitive function, developmental delays, and behavioral problems. A person may also develop lung cancer, kidney disease, and gastrointestinal illness.

Chlorine Compound

The final component that makes up Ashcrete is chlorine compound. As a result of its harmful effects, chlorine is considered to be an undesired compound. It is even said that Pliny Fisk, when he created Ashcrete for the first time, tried to find a substitute for the chloride compound in this product (Lerner, n.d). The migration of chloride into concrete accelerates the corrosion of embedded steel in concrete, and specific chloride cations are capable of corroding steel in concrete to varying degrees. In comparison to sodium chloride and potassium chloride, calcium chloride has a more detrimental effect (Zhu et al, 2008). To maximize the production and usability of Ashcrete, the chloride content must be reduced as much as possible. Nevertheless, Ashcrete has certain properties that prevent chloride from penetrating into the material.

Applications of Ashcrete

Despite some disadvantages, Ashcrete remains a better environmentally-friendly alternative to concrete. Living in a house built on coal infrastructure is still a matter of debate as to whether or not it is safe. The disadvantages and advantages of components of Ashcrete and their potential threat on the environment limits its application. In recent years, there has been an increase in its application in several settings. One example is the application of Ashcrete in marine structures. In many places, Ashcrete is already available in the market. While fly ash may not be found everywhere in the world, the use of Ashcrete is unlikely to be implemented in all countries across the globe. However, it can still create a significant environmental impact in places that do have access to fly ash. In order to ensure a stable supply of electric power, Japanese thermal power plants have diversified their fuel sources since the petroleum crisis (Ishikawa, 2007). The coal-fired thermal power generation equipment currently occupies 16% of the total power generation equipment, and it is estimated that this ratio will remain unchanged in the future (Ishikawa, 2007). In Japan, fly ash used to be a waste and nuisance to be disposed of by dumping in shallow waters more than ten years research and tests on Ashcrete has enabled the material to be officially qualified for major public construction projects such as fishing ground construction. Since 1980, artificial reefs made of Ashcrete have been constructed and installed at a number of locations in Japan after passing all the safety requirements set forth by the Fisher Agency of the Japanese Government (Ishikawa, 2007). In a series of tests, Ashcrete had proven to provide compressive strength, resistance to seawater, and safety for marine life. A study by Suzuku claimed that the addition of appropriate activators such as sodium chloride curing in artificial sea water (compared to tap water) will greatly enhance both initial and long term strength (Suzuku, 1995). Each year, the electric company in Japan produces approximately eight million tons of coal ash. Fly ash constitutes approximately 90 percent of the coal ash discharged in the country, with the remaining ten percent made up of clinker ash. Among various technologies that are being developed for effective use of fly ash, the use as an admixture for concrete is considered the most effective (Suzuku, 1995).

Conclusion

Concrete has a significant environmental impact, going further than releasing a large amount of carbon dioxide into the environment owing mainly to its cement production. However, the demand for concrete is growing much faster than the demand for steel or wood in comparison to previous few years. To satisfy the need for concrete while reducing the impact its production has on the environment, alternatives such as Ashcrete can be implemented. Ashcrete could ultimately prove to be a viable alternative owing to its greater strength and lower carbon footprint. Ashcrete has not yet gained prominence in construction outside of the U.S. However, dam builders have increasingly preferred this material due to its strength and resistance to water damage. Rather than letting the main components of Ashcrete, such as fly ash, end up in landfills where it can have its own negative environmental impact, they can be used for a more sustainable purpose.

References

Andrade, L. B., Rocha, J. C., & Cheriaf, M. (2009). Influence of coal bottom ash as fine
aggregate on fresh properties of concrete. Construction and Building Materials, 23(2), 609–614. https://doi.org/10.1016/J.CONBUILDMAT.2008.05.003

Beyondhomes (2020). Ashcrete VS Concrete. Beyondhomes Home Interior Design. Retrieved
January 14, 2023 from https://www.beyondhomes.ca/ashcrete-vs-concrete/

Chun, L. B., Sung, K. J., Sang, K. T., & Chae, S. T. (2008, November). A study on the fundamental
properties of concrete incorporating pond-ash in Korea. In The 3rd ACF international conference-ACF/VCA (pp. 401-408)

Coal Fly Ash - Material Description - User Guidelines for Waste and Byproduct Materials in Pavement Construction - FHWA-RD-97-148. (2016). Federal Highway Administration. Retrieved January 15, 2023, from https://www.fhwa.dot.gov/publications/research/infrastructure/structures/97148/cfa51.cfm

Davraz, M. (2010). The effects of boron compounds on the properties of cementitious
composites. Science and Engineering of Composite Materials, 17(1), 1–17. https://doi.org/10.1515/SECM.2010.17.1.1/MACHINEREADABLECITATION/RIS

Energy Education. Coal fired power plant. (n.d.). Retrieved January 15, 2023, from
https://energyeducation.ca/encyclopedia/Coal_fired_power_plant

Facilitator, C. An overview of fly ash; classification, advantage, and utlization. (2022). Constro
Facilitator. Retrieved January 15, 2023, from https://constrofacilitator.com/an-overview-of-fly-ash-classification-advantage-and-utlization/

Gomaa, E., Sargon, S., Kashosi, C., & ElGawady, M. (2017). Fresh properties and compressive
strength of high calcium alkali activated fly ash mortar. Journal of King Saud University - Engineering Sciences, 29(4), 356–364. https://doi.org/10.1016/J.JKSUES.2017.06.001

Ishikawa, Y. (2007). Research on the Quality Distribution of Jis Type-II flY ash in Japan. World
of Coal Ash. https://p2infohouse.org/ref/45/44777.pdf

Kim, H. K., & Lee, H. K. (2011). Use of power plant bottom ash as fine and coarse aggregates in
high-strength concrete. Construction and Building Materials, 25(2), 1115–1122. https://doi.org/10.1016/J.CONBUILDMAT.2010.06.065

Lerner, S. Eco-Pioneers: Practical Visionaries Solving Today's Environmental Problems. (n.d.).
Retrieved January 30, 2023, from https://www.washingtonpost.com/wp-srv/style/longterm/books/chap1/ecopioneers.htm

MaterialDistrict. Taking the 'con' out of concrete. (2019). Retrieved January 30, 2023, from
https://materialdistrict.com/article/con-out-of-concrete/

Material Insights. (n.d.). Retrieved January 15, 2023, from
https://www.material-insights.org/materials/boron-borates/

Moravej, M. Borates in Construction: Maximizing the Value of Essential Materials | U.S. Borax.
(2022). Retrieved January 15, 2023, from https://www.borax.com/news-events/january-2022/borates-in-construction

Nair, M. (2021). Alternative Material: AshCrete. Rethinking the Future. Retrieved January 14,
2023 from https://www.re-thinkingthefuture.com/materials-construction/a4497-alternative-material-ashcrete/

Singh, M., & Siddique, R. (2013). Effect of coal bottom ash as partial replacement of sand on
properties of concrete. Resources, Conservation and Recycling, 72, 20–32. https://doi.org/10.1016/J.RESCONREC.2012.12.006

Suzuki, T. (1995). Chemistry and Ecology Application of High-Volume Fly Ash Concrete to
Marine Structures. https://doi.org/10.1080/02757549508037682

What is Ashcrete? - The Constructor. (n.d.). Retrieved January 15, 2023, from https://theconstructor.org/building/ashcrete/565323/

What does a pozzolan do in the concrete? (n.d.). American Concrete Institute. Retrieved January
15, 2023, from https://www.concrete.org/tools/frequentlyaskedquestions.aspx?faqid=689

Yüksel, I., & Genç, A. (2007). Properties of concrete containing nonground ash and slag as fine
aggregate. ACI Materials Journal, 104(4), 397–403. https://doi.org/10.14359/18829

Zhu, F., Takaoka, M., Shiota, K., Oshita, K., and Kitajima, Y. (2008). Chloride Chemical Form in Various Types of Fly Ash. 319–1195. https://doi.org/10.1021/es7031168

Chapter 4: Timbercrete
by Sarah Mansoor

Introduction

Timbercrete is a revolutionary construction material that blends sawdust and sand to create a sustainable and eco-friendly alternative to traditional concrete (Hamakareem, 2022). It utilizes waste products and reduces energy consumption during production, promoting environmental conservation (Timbercrete Pty. Ltd., 2015). With the ability to be molded in various sizes, colors, shapes, and textures, Timbercrete is suitable for a wide range of building projects, including homes and residential buildings (Architecture and Design, 2012). Timbercrete is able to sequester carbon and store it in the building, offsetting emissions from polluting vehicles (The Uptide, 2021). It is also lighter and has exceptional thermal insulation properties, making it a perfect choice for sustainable building projects (Kumar, Kumar, Uddayappan, Raj, & Babu, 2019). Timbercrete is an Australian invention, with several international patents and trademark protection in most of the world's population (Timbercrete Pty. Ltd., 2015). It is also the only structural brick or block product on the Australian market that traps carbon which would normally end up as greenhouse gases in our atmosphere (Timbercrete Pty. Ltd., 2015).

How is Timbercrete Made?

The process of creating Timbercrete starts by blending sawdust and sand together in a mixer (Timbercrete Pty. Ltd., 2015). Then, cement is added to the mixture and blended until it is evenly distributed. Once the mixture has a consistent color, water is slowly and evenly added and blended again (Timbercrete Pty. Ltd., 2015). This ensures that the Timbercrete mixture is homogeneous, resulting in a sturdy and reliable construction material. Timbercrete is made of a unique blend of cellulose (timber waste), cement, sand, binders, and other materials

(Timbercrete Pty. Ltd., 2015). The primary ingredient in Timbercrete is timber waste, such as sawdust or recycled timber from discarded pallets (Timbercrete Pty. Ltd., 2015). This material is obtained from sawmill waste from plantation timbers, rather than cutting down trees specifically for Timbercrete production (Timbercrete Pty. Ltd., 2015). Other ingredients include sand, which is carefully selected to maximize its load-bearing capacity and minimize water ingress (Timbercrete Pty. Ltd., 2015), and cement binders, such as Portland cement or other cementitious material, along with a special nontoxic "deflocculate" and other products that improve density and cement performance (Timbercrete Pty. Ltd., 2015). These products also function as a waterproofing agent.

Features of Timbercrete

According to Hamakareem (2022), Timbercrete has several notable features, including its density (which can range from 900 Kg/m3 to 1500 Kg/m3), load-bearing capacity (varying between 5 MPa and 15 MPa), bulletproof capabilities (as no bullet has been able to penetrate through a 200 mm brick), insulation value (with an R-2.5 per 25 mm of thickness), and fire resistance. Additionally, Timbercrete is 2.5 times lighter than concrete and clay, making it easier to handle and transport.

In terms of sound transmission and absorption, Timbercrete Pty. Ltd. (2015) reports that Timbercrete has a good performance in preventing sound transmission, with a "weighted sound absorption coefficient" of 0.20 and a noise reduction coefficient (NRC) of 0.15, which is higher than typical clay and concrete products (NRC of 0.04). Timbercrete also provides better resistance to airborne sound transmission than aerated concrete, while providing less deflecting sound than higher density clay and concrete.

Furthermore, Timbercrete Pty. Ltd. (2015) states that the formula for Timbercrete can be altered to achieve specific engineering requirements, and that load-bearing typically ranges from 5 MPA to 15 MPA or potentially greater if necessary. At a 5 MPA load-bearing capacity, a single standard 400 long X 200 mm thick Timbercrete block can support up to 40 tonnes. Tradesmen also prefer working with Timbercrete because it is lightweight (at $1,100 kg/m3$) and can be nailed and screwed without the need for pre-drilling, similar to traditional timber.

Benefits of Timbercrete

The use of Timbercrete has several benefits, including its ability to trap carbon dioxide and promote environmental conservation (Hamakareem, 2022). Timbercrete has a lower embodied energy compared to clay bricks (Hamakareem, 2022), making it a more sustainable option for construction. Timbercrete has a higher insulation value (R) than conventional solid masonry bricks, blocks, and panels (Hamakareem, 2022), making it an energy-efficient option for buildings. Timbercrete is also a thermal mass material that absorbs and stores thermal energy, making it ideal for high thermal insulation projects such as homes and residential buildings (Hamakareem, 2022). Timbercrete blocks and panels can easily be screwed and nailed without losing conventional masonry advantages (Hamakareem, 2022), making it a versatile and flexible material for a wide range of building projects. The units of Timbercrete are lighter and larger, making it easy to handle and quickening the construction process (Hamakareem, 2022). Furthermore, the materials used in Timbercrete are locally available, reducing transportation costs and overall cost (Hamakareem, 2022), making it a cost-effective and sustainable alternative to traditional concrete.

Timbercrete can be used in the construction of sound barriers and retaining walls. The wood fibers in the mix can improve the sound-absorption properties of the material, making it an effective solution for reducing noise pollution in urban areas. Timbercrete's ability to be poured or cast into various forms also allows it to be used in the construction of retaining walls and other structures that require a strong and durable material (Broughton et al., 2016).

Timbercrete can also be used in the construction of low-cost housing. The material is relatively inexpensive to produce and can be used to create affordable and sustainable housing for low-income communities (Broughton et al., 2016).

Another advantage of Timbercrete is its improved thermal insulation properties. The wood fibers in the mix provide insulation, which can help to reduce energy costs for heating and cooling buildings. This can be especially beneficial in colder climates where traditional concrete buildings may require significant heating to maintain comfortable temperatures (Broughton et al., 2016).

Timbercrete also has improved fire resistance compared to traditional wood-based building materials. The cement in the mix provides fire resistance, making Timbercrete a suitable alternative for use in buildings where fire resistance is a concern (Broughton et al., 2016).

Applications of Timbercrete

Timbercrete is a versatile and eco-friendly building material with a wide range of applications. Some of the most common uses of Timbercrete include residential buildings, industrial and commercial buildings, landscaping design, acoustic barriers for highways, and cladding panels (Hamakareem, 2022).

In residential construction, Timbercrete's high thermal insulation and fire resistance properties make it an ideal material for building homes and residential buildings. Its ability to trap carbon dioxide and lower embodied energy also make it a sustainable and environmentally-friendly alternative to traditional concrete (Hamakareem, 2022).

In industrial and commercial construction, Timbercrete's high load-bearing capacity and durability make it an ideal material for building industrial and commercial buildings. Its eco-friendly properties also make it a sustainable choice for commercial buildings (Hamakareem, 2022).

In landscaping design, Timbercrete can be molded into various shapes and sizes, making it a versatile material for creating garden beds, retaining walls, water features, and other landscaping features (Hamakareem, 2022).

In acoustic barriers for highways, Timbercrete's high sound insulation properties make it an ideal material for reducing the transmission of noise pollution and improving the quality of life for residents living near highways (Hamakareem, 2022).

In multi-story apartments, Timbercrete's high fire resistance and sound insulation properties make it an ideal material for building walls, improving the safety and comfort of residents (Hamakareem, 2022).

In cladding panels, Timbercrete's ability to be molded into various sizes, colors, shapes, and textures make it an ideal material for creating aesthetic and durable cladding for buildings (Hamakareem, 2022). Timbercrete can be used in other building projects such as bridges, roads, and other infrastructure projects. Its versatility and eco-friendly properties make it a suitable material for sustainable building projects. With further research and development, Timbercrete's potential applications in the construction industry could be even broader.

Timbercrete also has excellent insulation properties, making it an ideal material for energy-efficient construction. It has a high thermal mass, which helps to regulate indoor temperatures and reduce energy consumption for heating and cooling. This makes Timbercrete a great choice for passive solar design, where the sun's energy is harnessed to heat and cool the building naturally.

Timbercrete is also suitable for use in sound insulation and acoustic applications. Its high density and unique composition help to reduce sound transmission, making it an ideal choice for use in sound-sensitive areas such as recording studios and music rooms.

In addition to its structural, insulation, fire resistance and sound insulation properties, Timbercrete has many other advantages. It is easy to handle, can be shaped and cut to any size or shape, and is easy to finish. It is also less expensive than traditional concrete, and can be used to create stunning architectural features.

In summary, Timbercrete offers a wide range of applications, from foundations, walls and floors, to fire-rated walls, sound insulation and acoustic applications. Its eco-friendly, sustainable, and versatile properties make it a great choice for residential, commercial and infrastructure projects. Furthermore, it is cost-effective and easy to handle, making it a perfect choice for many projects.

Environmental Impacts of Timbercrete

The construction industry has long been known for its negative impact on the environment, due to its heavy pollution and resource consumption. However, sustainable architecture practices have led to the development of materials like Timbercrete, which offer a more environmentally-friendly alternative without compromising on aesthetics or quality. Timbercrete is a low-carbon footprint construction material that can help to reduce the environmental impact of the construction industry (The UpTide, 2021). It is made from recycled timber waste, such as sawdust, and has significantly lower embodied energy compared to traditional concrete (Timbercrete An Introduction, 2015). Additionally, Timbercrete acts as a carbon trap, preserving cellulose waste within a concrete tomb and preventing the release of greenhouse gasses (Timbercrete An Introduction, 2015). Timbercrete does not require artificial or man-made drying processes, and its production process consumes less equipment and energy than other traditional brick and block making systems (Timbercrete An Introduction, 2015). Timbercrete also utilizes locally sourced raw materials wherever possible, further reducing its environmental impact (Timbercrete An Introduction, 2015). Overall, Timbercrete offers a sustainable and eco-friendly alternative to traditional concrete that can help create a more sustainable future for the construction industry.

Timbercrete vs Concrete

Timbercrete is a sustainable and eco-friendly alternative to traditional concrete, with a focus on achieving the same level of strength and stability while using a thinner composition (The Uptide, 2021). This allows for increased usable space within walls while still maintaining excellent insulation properties. Timbercrete's minimal material usage and reduced waste result in a lower environmental impact compared to traditional concrete. The use of locally sourced wood in its production, when possible, reduces the environmental impact of transportation and the importation of building materials (The Uptide, 2021).

One of the main advantages of Timbercrete is its cost-effectiveness, as the use of waste products such as sawdust in its production can lower the overall cost of the material, making it more accessible to the average consumer. Despite being a relatively new construction material, Timbercrete has been gaining popularity in recent years (The Uptide, 2021).

Timbercrete's environmentally friendly properties are further highlighted by its unique density and matrix, which achieve improved acoustic and thermal insulation compared to conventional bricks (Timbercrete An Introduction 2015). Its unique resilience and good load-bearing capacity make it easy to work with, allowing for fast and easy construction. The dry density of Timbercrete can be altered to suit specific requirements, ranging from 900 kg/m^3 to 1500 kg/m^3 (Timbercrete An Introduction 2015). Standard Timbercrete has a density similar to water and hardwood timber, making it lightweight and easy to handle (Timbercrete An Introduction 2015).

Developing Countries

In developing countries, construction is a top priority as there is a constant demand for both residential and non-residential buildings (Kumar et al., 2019). One way to reduce the cost of construction is to use alternative materials that are readily available, such as agricultural wastes and disposed materials (Kumar et al., 2019). One such material that can be used as a replacement is sawdust, which can easily be obtained from furniture and wood industries (Kumar et al., 2019). Utilizing this waste product instead of disposing of it by burning can also have positive environmental impacts (Kumar et al., 2019).

The demand for concrete, a primary component in construction, is increasing globally (Kumar et al., 2019). However, the use of fine aggregate in concrete can lead to scarcity and an increase in demand for sand (Kumar et al., 2019). In various research studies, sawdust has been used as an alternative to fine aggregate in the production of solid blocks (Kumar et al., 2019). However, converting sawdust into ashes by burning it can lead to the emission of harmful gasses such as carbon dioxide and carbon monoxide (Kumar et al., 2019). These excess emissions can lead to various health hazards and contribute to a larger carbon footprint (Kumar et al., 2019).

Timbercrete can be a solution to this problem as it allows for the use of sawdust as a primary ingredient without the need for burning (Kumar et al., 2019). This not only reduces waste for the timber industry but also offers a cost-effective and environmentally friendly alternative to traditional concrete (Kumar et al., 2019).

Limitations of Timbercrete

Despite its potential as a sustainable construction material, Timbercrete has certain limitations that must be taken into consideration. According to The Uptide (2021), one of the main limitations is the lack of standardized mixing processes, which can result in poor quality control. Using waste wood as a source material for Timbercrete may pose health risks if mixed with other chemicals such as formaldehyde. The strength and stability of Timbercrete has not yet been fully established, and further testing is required to fully understand its capabilities (The Uptide, 2021). It is important to note that Timbercrete is not a perfect solution, and it is not completely pollution-free. As urbanization and development continue to increase, the demand for traditional concrete will likely rise. Therefore, it is crucial to take steps to reduce carbon emissions as much as possible, even if it is not possible to eliminate them entirely (The Uptide, 2021).

One of the limitations of Timbercrete is that it can be affected by moisture. The wood fibers in the mix can absorb water, which can cause the material to warp or rot over time. To prevent this, Timbercrete should be protected from moisture by using appropriate coatings or sealers (Broughton et al., 2016).

It is also worth noting that, while Timbercrete has the potential to replace traditional concrete in many applications, it may not be suitable for all projects. For example, large-scale infrastructure projects such as bridges or highways may require the high compressive strength of traditional concrete. However, for smaller-scale projects and where structural integrity is not as crucial, Timbercrete can be a more sustainable and cost-effective alternative.

Research Study on Timbercrete

A research study conducted by Kumar et al. (2019) in India aimed to investigate the properties of Timbercrete blocks by using sawdust as a partial replacement for fine aggregate in the concrete mixture. The study used a range of sawdust replacement levels, from 10% to 50% by weight, and evaluated the blocks' compressive strength and water absorption after 28 days of curing. The study found that the use of sawdust in concrete allowed for the disposal of water and made the blocks lighter in weight, which could lead to more efficient construction. Timbercrete was found to be an environmentally-sensitive building material that can trap carbon and reduce greenhouse gas emissions. The study concluded that sawdust can be effectively used as a replacement for fine aggregate in Timbercrete, with the optimal replacement level found to be around 30%.

The study also found that the water absorption of timbercrete blocks was significantly lower than that of traditional concrete blocks, indicating that timbercrete blocks are more water-resistant (Kumar et al., 2019). This is an important consideration for construction in areas prone to flooding or heavy rainfall.

The study found that the use of sawdust as a replacement for fine aggregate in timbercrete blocks resulted in a reduction in weight of the blocks, making them easier to handle and transport during construction (Kumar et al., 2019). This also has the added benefit of reducing the load on the structure, allowing for more efficient and cost-effective construction.

The researchers also found that the use of sawdust in timbercrete blocks did not negatively impact the overall strength of the blocks. In fact, the compressive strength of the blocks made with sawdust replacement was found to be comparable to that of traditional concrete blocks (Kumar et al., 2019).

Overall, the study conducted by the International Research Journal of Engineering and Technology (IRJET) in India provides strong evidence that Timbercrete is a viable and sustainable alternative to traditional concrete, and that the use of sawdust as a replacement for fine aggregate in timbercrete blocks can result in significant benefits in terms of cost, strength, and environmental impact (Kumar et al., 2019).

Conclusion

In conclusion, Timbercrete is an innovative and environmentally friendly construction material that has a wide range of benefits and can be used for a variety of building projects. Its unique properties, such as its density, load-bearing capacity, bulletproof capabilities, insulation properties, fire resistance, durability, sound insulation and environmental benefits make it a suitable and sustainable alternative to traditional concrete. Timbercrete's ability to trap carbon dioxide, lower embodied energy, higher insulation value, thermal mass properties, versatility, ease of handling, and cost-effectiveness make it an ideal material for a wide range of building projects. Further research and development is needed to fully understand the potential of Timbercrete and its potential applications in the construction industry. Additionally, the research study conducted by Kumar et al. (2019) on the strength of Timbercrete blocks has shown that sawdust can be used as a partial replacement for fine aggregate, resulting in a reduction in the demand for river sand and a reduction in the weight of the blocks. This can lead to more efficient construction, as the load on the structure is reduced, allowing for more floors to be added to a building. The study also found that Timbercrete blocks made with sawdust had a compressive strength of up to 90% after 28 days of curing. Overall, the research supports the conclusion that Timbercrete is a promising sustainable alternative to traditional concrete and offers many benefits to the environment and the construction industry. It is important to continue research and development to fully understand the potential of Timbercrete and its capabilities.

References

Architecture and Design (2012, March 18). Timbercrete bricks, blocks and panels. Retrieved January 13, 2023, from https://www.architectureanddesign.com.au/suppliers/timbercrete/timbercrete-bricks-blocks-and-panels#

Broughton, R., Treloar, G., & Broughton, J. (2016). Timbercrete: A sustainable alternative to traditional building materials. Journal of Sustainable Development, 9(4), 48-54.

Hamakareem, M. I. (2022, July 13). Timbercrete: Components, advantages, and applications. The Constructor. Retrieved January 13, 2023, from https://theconstructor.org/building/timbercrete-components-advantages-applications/565251/

Kumar, M. V., Kumar, L. V., Uddayappan, S. S., Raj, R. Y., & Babu, S. S. (2019). Study on Strength of Timbercrete Blocks. International Research Journal of Engineering and Technology (IRJET). Retrieved January 2023, from https://www.irjet.net/archives/V6/i3/IRJET-V6I3649.pdf

Timbercrete Pty. Ltd. (2015). Timbercrete An Introduction. Timbercrete. Retrieved January 2023, from https://timbercrete.com.au/wp-content/uploads/2021/08/A-Introduction_to_Timbercrete-V4.pdf

The Uptide. (2021, October 20). Timbercrete explained: Green Construction Guide. The Uptide. Retrieved January 13, 2023, from https://www.theuptide.com/what-is-timbercrete/

Chapter 5: Aircrete
by David Henneberg

What is Aircrete?

"Concrete is one of the most widely used, human-made materials on the planet. Some estimates state that we have made enough concrete to cover the Earth's entire land surface in a thin coat of cement." (Roberts, 2020) That is a direct quote pulled from Aircrete Guide: Everything You Need To Know. It may come as no surprise that concrete is a fundamental building block for the infrastructure of modern civilization. From buildings, roads, sidewalks to bridges, concrete is used in almost every construction project imaginable (and that list is by no means a comprehensive list of all its applications). Concrete itself is a "composite material composed of fine and coarse aggregate bonded together with a cement paste that hardens over time." ("Concrete", n.d.) From a quick Wikipedia search one can see that its "usage worldwide, ton for ton, is twice that of steel, wood, plastics, and aluminum combined." ("Concrete", n.d.) One issue that the world is seeing because of this mass amount of production is an increase in greenhouse gas emissions. In fact, "the production process for cement produces large volumes of greenhouse gas emissions, leading to net 8% of global emissions." ("Concrete", n.d.) One of the practical solutions of Aircrete is that it is more biodegradable and emits less toxic material when being processed than its predecessor. Although the environmental impacts of Aircrete vs concrete are important (which will be covered later on in the chapter), there are many other areas to look at first.

So what is Aircrete and how does it differ from concrete? "Aircrete is simply concrete with bubbles. Regular concrete that we use for our roads, basements, and foundations traditionally is made from Portland Cement. This combination hardens into highly dense material with impressive compressive strength." (Roberts, 2020) The process for creating Aircrete differs. "Aircrete reduces or eliminates the traditionally used aggregate. Instead of including gravel or other coarse aggregate types, aircrete relies on incorporating premade foam pieces that essentially add bubbles of airspace within the concrete mix." (Roberts, 2020) Another way to think of Aircrete is as a "mix of water, foaming agent, and cement. The foaming agent creates tiny air bubbles that, when evenly dispersed, provide many benefits." (Nature of Home, n.d.) The ratios that are used with water, foaming agent and cement may differ. The more foaming agent used in the mix will create a lighter material, but it will also offer less compression strength. "The target amount of foaming agent depends on the intended application for aircrete." (Nature of Home, n.d.) This makes complete sense as not all projects require the same level of durability. Building a shed in a backyard would require less strength than would be required of a highway overpass, for example. It is common for builders to use Aircrete in tandem with traditional concrete if they need to make the Aircrete stronger. It is important to note that Aircrete can be known as "cellular concrete, foam concrete, foamcrete, lightweight concrete, (and) aerated concrete." (DIYAircrete, n.d.) While these different names seem to be quite different, the fundamental ingredients to produce the product are virtually identical. Depending on the ratios of the mixes, or where it is mixed or who it is being used by, one might call the material by a different name.

There are many reasons Aircrete is becoming increasingly popular for construction projects and do-it-yourself builders alike – most of which will be covered later in this chapter. The reason Aircrete was invented was "to develop a simple, inexpensive, do-it-yourself method." (DIYAircrete, n.d.) The inventor of Aircrete is Hajja Gibran who created Domegaia, a company that builds striking, eco-friendly homes with Aircrete. These homes are spherical, and Hajja looks at his projects as a solution to affordable housing, especially in low-income areas of the world with tough weather conditions. Hajja says that "'domes are the strongest structures … they can withstand the forces of winds and earthquakes … cement is fireproof (and) waterproof, so it is not going to be destroyed by fire, water, and floods.'" (Thompson, 2021)

"For years, aircrete was only created commercially with the use of heavy-duty equipment. However, when Domegaia founder, Hajjar Gibran, discovered the properties of aircrete, he invented special foam generators called the Little Dragon and the DragonXL to make it available to everyone." (Domegaia, n.d.) Hajjar Gibran has a vision of creating homes that fit more naturally into their surroundings. The rest of this chapter will dive deeper into how it is made and the specifics on its applications.

How it is Made

What better place to get information than from the company that invented aircrete? As aforementioned Aircrete is made up of a mixture of water, cement powder and water-based foam. The following is a more technical description of the material: "Aircrete foam is produced by agitating a foaming agent with compressed air. A good quality degreasing dish detergent can be used as a foaming agent. Not just any foam works for aircrete, it is a high quality, specialized formula with specific density ... even after Aircrete becomes solid, it's easy to chisel, saw, cut holes into, craft into irregular shapes and so much more ... Aircrete is easy to make but much like baking a cake, you have to get the recipe right. The quality and density of the foam is important. Use an accurate postal or kitchen scale to check the weight of your foam. It should be between 90-100 grams/liter." (Domegaia, n.d.) This is interesting to note if one is wanting to try and create Aircrete at home. There seems to be a certain level of improvisation, trial and error to get the mix correct. Depending on the working conditions one is currently operating in could determine the right mix. The point is, a certain mix of material might be right for a person constructing a house in the Caribbean in a hot and humid environment, while another mix might be right for a person working in a dry and cold environment in the northern hemisphere.

To continue with the theme of improvisation it is possible to "use good quality dish detergent to make the foam with Little Dragon, our continuous foam generator. Look for a high foaming degreaser detergent. We tested Seventh Generation Natural Dish Liquid, Dawn Ultra and Safeway Home concentrate. They all produced adequate foam diluted 40/1 with water." (Domegaia, n.d.) This just goes to show how new Aircrete is as a material. There are a variety of ways to get a similar end product.

The technical steps include mixing one 94 lb bag of cement with 6 gallons (23 liters) of water. Put all the water in your container first and add the cement while you are mixing to avoid clumping. When the cement and water are well mixed, turn on the Little Dragon (foam generator) and add foam to the mixture. Add enough foam to make a total of 45 gallons (170 liters) of Aircrete." (Domegaia, n.d.)

Again, there seems to be no single "correct" way to make Aircrete. Anyone that wishes to create Aircrete should try it on their own and figure out what mix of materials (concrete, foaming agent) is required for their own specific projects.

Qualities of Aircrete

"Aircrete is fireproof, insect proof and unharmed by moisture – it will not rot or decay." (Domegaia, n.d.) Since Aircrete is less expensive than a lot of building materials used in houses, it might be right for a person living in a place with termites. Termites eat wood and destroy the foundations of houses. If a foundation is made with Aircrete it would be unaffected by these pests.

"Aircrete has good compression strength but poor tensile strength." (Domegaia, n.d.) This basically means that Aircrete is a lot stronger in forms that are curved. As a flat surface, Aircrete is much weaker. "All other factors being equal, the more the curvature, the stronger the form. Conversely, flatter surfaces create weaker forms." (Domegaia, n.d.) The original designers of Aircrete point out the weaknesses in Aircrete here. The Domegaia company designs structures under high levels of compression, which are circular in shape. Think about hobbit style homes, or the igloo style homes from Star Wars, and that is how Domegaia designs homes. This is due to the poor tensile strength of Aircrete, and so its application for curved structures is much more practical than flat structures.

Hajjar Gibran says "when the forces acting upon a compression shell are understood and followed, we can build elegant structures of magnificent integrity. The art of designing equilibrium into a structure can create poetic forms, composed of harmonic geometry, expressing a graceful dance with the forces of nature.' (Domegaia, n.d.) It is recommended to check out the exceptional building designs on Domegaia's website using Aircrete. That being said, flat surfaces, or walls, have been created successfully with Aircrete. There is a need for more reinforcing materials, however.

Potential Applications

"On its own, Aircrete is breakable, however when you surround it by layers, you'll have a strong structure." (Domegaia, n.d.) On Domegaia's website they show detailed diagrams and blueprints for their structures with Aircrete. Aircrete is meant to be used in conjunction with other materials, or layers, in order to be functional. "Aircrete needs to be poured into a form and allowed to harden overnight. It makes good foundation footings, slabs and sub-floors. Lay plastic down to keep water from evaporating so it cures thoroughly. In dry climates sprinkle it with water to keep it wet for a few days to help the curing process." (Domegaia, n.d.) Keep in mind the importance of layering with Aircrete, as it will inform the integrity of the structure built.

"It offers good thermal and acoustic insulation. Unlike concrete which is hard, heavy, cold and difficult to work with; Aircrete is easy to work with." (Domegaia, n.d.) The lightweight factor of Aircrete is a seriously wonderful quality. Those that have worked with typical cement or concrete understand how heavy it is and difficult it is to rework or reshape. Aircrete allows the user to build with a material that is strong but also much more malleable.

"It hardens overnight and can be cut, carved, drilled and shaped with wood-working tools. It accepts nails, screws and is easily repaired. It continues to harden over time and makes excellent foundations, subfloors, building blocks, walls, dome arches or whatever. It can be molded or formed into practically any shape." (Domegaia, n.d.) The practical applications of Aircrete are endless, and as long as the material is used properly it will last a very long time.

Another application of Aircrete is that it "can be used with the 'cookie cutter' method to make building blocks. Make rectangular wooden Aircrete forms by joining the corners with door hinges that have removable pins for easy assembly and disassembly." (Domegaia, n.d.)

This allows for a whole swath of prefabricated systems to be put in place. Walls, roofs, and stairs (to provide just a few examples) could be made ahead of time and stored. Since Aircrete is so much lighter than its predecessor, it is easily transferred to construction sites as needed.

Current Issues Surrounding Aircrete

In North America Aircrete has not become common use in the building industry but "Europe and Asia Aircrete homes are very common and even more so now." (Landzero, n.d.) This is due to different regulations and building standards - essentially the status quo here in North America. Aircrete would be a lot more common if there were not such stringent codes to be followed in building projects in most areas of North America. In order to use Aircrete in any sort of structure a permit is required, otherwise the structure would not be insurable. This is a typical case of red-tape holding back innovation.

Another issue in using Aircrete is that it needs to be self-made in North America. There are places that manufacture the product but they are few and far between, so in all likelihood the builder would need to travel long distances to get their hands on it. Luckily, the method for creating Aircrete is not too complicated, and with a little trial and error one can easily make it themselves.

It would seem that Aircrete takes a long time to cure, and it can be in a weak and crumbly state until it matures. This can take up to one month. It is especially important to be reminded that Aircrete has low load-bearing strength, and should not be used for driveways, garage floors or patios.

Also, "if you're purchasing pre-manufactured aircrete, be aware that some manufacturers use foam products containing harmful chemicals. However, it is entirely possible to make aircrete with non-toxic natural resources." (Nature of Home, 2022) If outsourcing, always use caution, as it can be cheaper to build with toxic materials rather than safe, environmentally friendly materials.

Aircrete 61

Environmental Implications

Reading on the MPA Masonry website, one can read that "Aircrete is a durable building material with a very long life expectancy (well in excess of the Green Guide to Housing Specifications 60 year building life). It requires little maintenance over its lifespan and is a good alternative to timber framed buildings. Aircrete is a durable material that does not give off any harmful substances or require preservatives to maintain performance. It does not rot or burn. It is resistant to sulfates and the effects of freeze/thaw cycles and it cannot be attacked by pests such as termites or vermin." (MPA, n.d.) All of these attributes of Aircrete make it an environmentally viable option in comparison to traditional concrete. Traditional concrete as a material is responsible for more greenhouse gas emissions than any other building material - even more than plastic! Concrete has been an overlooked and ignored component of the greenhouse gas effect, and Aircrete offers a hopeful solution to the problem. Although Aircrete cannot replace concrete in every area due to the differences in tensile strength, there are many applications that Aircrete could replace the traditional building material (i.e. lighter projects that require less tensile strength.

Since the injection of air into concrete is the fundamental difference between traditional concrete and Aircrete, there is less material going into its creation. Less material equals less material consumption which equals less energy consumption. "When making aerated concrete, cement is expanded six times its original volume with air reducing the carbon footprint." (BuilderSpace, n.d.) On top of this, "the disposal of aerated concrete does not bring any harm to the environment." (BuilderSpace, n.d.) This contrasts with traditional concrete greatly, which has a variety of harmful chemicals and toxic materials (like hexavalent chromium, or chromate, which can produce ulcers in the human body with long periods of exposure).

Economical Factors

Aerated concrete as a product has a less expensive list of materials required for its creation. "Aircrete is a high-quality, low-cost material that eliminates the need for aggregates such as gravel, sand, and rock. Conversely, concrete is a composite material that employs coat aggregates for strength making it more expensive than aerated concrete." (BuilderSpace, n.d.) The simple fact of the matter is that Aircrete is a cheaper building material than concrete. This does not mean that it is less capable - on the contrary - in previous sections of this chapter describing the possible applications of Aircrete one can see that if used correctly and for the right project, Aircrete can perform just as well as its more expensive counterpart.

A secondary economic factor is energy usage. Electricity use can become quite expensive for homeowners when their homes are poorly insulated. "Concrete, which is the most popular building material in the world, is not a good insulator because of its resistance to heat flow. Thus a concrete structure will not reduce electricity consumption..." (BuilderSpace, n.d.) Aircrete on the other hand has fantastic insulation qualities. According to BuilderSpace, "Aircrete offers excellent insulation effects and saves energy. Aerated concrete helps a homeowner save a considerable amount of money on bills throughout the year." (BuilderSpace, n.d.)

The rising costs of electricity have been felt around the world in recent years as more and more electricity is required. It is an often overlooked subject, but this trend is likely to increase as more and more parts of the world's energy economy rely on electricity. Electric cars are a prime example of this. As the transition from fossil fuels (i.e. gasoline) to electric vehicles continues to rise, the cost of electricity will likely skyrocket even higher. This is due to factors outside of the scope of this chapter. The main point is that becoming more conscious of efficiency will become increasingly important as we move into the future. Aircrete, being a more affordable building material with more robust insulation properties will have to become more common in order to keep homes and buildings affordable.

Chapter Summary

The fundamental takeaway is that Aircrete is a building material for the future, at least in North America, since Aircrete has been a prominent building material in Europe and Asia for well over a decade. When it comes to the status quo in building, North Americans are generally slower to change and react to new possibilities. The transition will happen for a number of reasons, where applicable. This is because the material is amazing. It is strong, malleable, moveable, environmentally friendly and economical. The only downside is that it has poor tensile strength, making its application more difficult for constructing flat surfaces. This pitfall is a fair trade-off considering all of the benefits.

Wherever Aircrete is found to be applicable, it will be. It will not just be a Do-It-Yourself material, either. It will be used in construction sites in home building as well as commercial building. There will be applications for its use on common infrastructure like highways and overpasses which has the potential for a revolutionary shift in greenhouse gas emissions. Remember, behind fossil fuels and livestock, cement is the number one polluter on the planet Earth.

References

Aircrete 101: The ultimate guide to how to make Aircrete and much more. Domegaia. (n.d.). Retrieved January 10, 2023, from https://domegaia.com/blogs/aircrete-dome-mastery/aircrete-101-the-ultimate-guide

Aircrete - everything you need to know. LandZero. (n.d.). Retrieved January 19, 2023, from https://www.landzero.com/blogs/we-love-land/aircrete-everything-you-need-to-know

Aircrete guide: Everything you need to know. Rise. (2021, July 10). Retrieved January 7, 2023, from https://www.buildwithrise.com/stories/aircrete-everything-you-need-to-know

BuilderSpace. (2021, June 25). Aircrete vs. concrete: Which is better? BuilderSpace. Retrieved January 22, 2023, from https://www.builderspace.com/aircrete-vs-concrete-which-is-better

Natureofhome. (2022, June 22). Aircrete: What it is, Pros & Cons - uses in homes/domes. Nature of Home. Retrieved January 9, 2023, from https://natureofhome.com/aircrete-what-it-is-dome-homes/

Sustainable block. MPA Precast. (n.d.). Retrieved January 20, 2023, from https://www.aircrete.co.uk/APA/Product-Information/Sustainable-Block.aspx

Thompson, D. (2021, April 19). This company is using 'Aircrete' to create remarkable, low-cost dome homes: 'A pretty indestructible building'. Retrieved January 9, 2023, from https://www.yahoo.com/news/innovative-dome-shaped-structures-may-132453122.html.

Wikimedia Foundation. (2023, January 1). Concrete. Wikipedia. Retrieved January 7, 2023, from https://en.wikipedia.org/wiki/Concrete

Chapter 6: Hempcrete
by Syed Rizvi

Climate change is a phenomenon that everyone is aware of, but often underestimate the concerning effects of. Researchers are looking for many different alternatives to combat it. However, world demand for resources is expected to double by 2050, where only 6% of materials are recycled (Belaïd, 2022). Cement and concrete are essential building blocks for our modern-day society. However, the building and construction sector contributes to 32% of global energy and 19% of greenhouse gas emissions (Yadav & Saini, 2022). Due to their abundant resources, easy operation, durability, and versatility, concrete is a widely used material. However, the intensive concrete use led by expanding real estate is contributing heavily towards climate change. Concrete is a synthetic rock, usually composed with sand, gravel, water, and cement (Belaïd, 2022). It is used in almost every building, transportation systems, water and wastewater systems, power systems, and many other necessary infrastructure for a society. Despite cement accounting for 10% of global concrete volume, it represents approximately 8% of greenhouse gas emissions (Belaïd, 2022). In the last 65 years, the demand for cement has increased tenfold in global commission, mostly used for concrete, but also mortars, plaster and blocks. However, cement production emits 1.5 billion tons of carbon dioxide approximately equal to 300 million European cars. "Cement production, for example, was responsible for ~7% (10.7 EJ) of the world's industrial energy consumption and 22% (2.2 Gt) of global GHG emissions arising from industrial processes in 2014" (Belaïd, 2022). Production of cement, the main ingredient in concrete, releases large amounts of carbon dioxide into the atmosphere in all of its stages. The first stage being the manufacturing stage, where raw materials are extracted, cooled and prepared. This stage uses the burning of fossil fuel to produce enough energy for the machinery and the cooling process. The next stage being the material stage, where producers of concrete, such as engineers, conduct performative criteria. The third stage involves the structural design and use, where architects, clients, designers, and construction companies mix the materials together in a manner most suitable for the target audience. The final is the end of the life and recycling stage, where concrete, cement, and other materials are put in recycling and/or demolition firms (Belaïd, 2022). However, the final stage involves power plant operations that deplete a huge quantity of energy, mostly coming from the burning of fossil fuel. Concrete production and use have a significant detrimental effect on the environment and cause climate change.

Concrete production, the most widely used building material worldwide, is responsible for about 8% of the world's CO_2 emissions. The main source of these emissions is the production of cement, which is an essential component of concrete. The process of calcination, which produces a sizable emission of CO_2, is used to make cement, which is mostly made of clay and limestone. The clay and limestone are heated at high temperatures during this process, which results in the production of CO_2 as a byproduct. Additionally, emissions are produced while raw materials and finished concrete products are being transported. Additionally, concrete is made using a lot of water and energy. The mining, crushing, and transportation of the raw materials necessary to manufacture concrete need a significant amount of energy. Water is also required in enormous amounts for mixing and grinding during the production of cement. Since there is still a high demand for concrete, these materials are becoming increasingly scarce. Although concrete contributes to climate change, there are steps that may be taken to decrease its harmful effects on the environment. One tactic is to use alternate materials or look for ways to use less concrete while building. For instance, using wood, bamboo, or other ecologically friendly materials during construction can reduce the amount of concrete needed because concrete use also contributes to climate change due to the effects of urban heat islands. Due to the absorption and retention of heat by concrete and other urban surfaces, metropolitan districts typically experience warmer temperatures than the neighboring rural areas. Increased energy demand for cooling as well as negative effects on the health of urban residents could emerge from this. Another method is to use other binders to create concrete. To reduce the quantity of cement used to manufacture concrete, scientists have been developing low-carbon binders including fly ash, slag, and silica fume. As a result, during the calcination process, less CO_2 is released.

By employing recycled materials to make concrete, the environmental impact can also be reduced. For instance, using recycled material instead of natural aggregate can reduce the amount of water and energy required for mining and shipping. In conclusion, concrete use and manufacture harm the environment and accelerate climate change. Concrete's detrimental effects on the environment can be mitigated in a number of methods, including the use of additional components, various binders, and recycled materials.It is imperative that individuals, organizations, and governments consider the environmental implications of concrete and take steps to decrease its contribution to climate change. Thus, many alternatives are proposed in order to tackle the issue at hand - introducing a new material called Hempcrete, also known as Lime-Hemp Concrete (Barbhuiya & Bhusan Das, 2022). Hemp is attracting more attention in the building and construction sector due to its properties of carbon-sequestering, higher biomass production and its insulation (Ahmed et al., 2022).

Firstly, many historians believe that hemp comes from central Asia and districts in India, especially in the Humalayan mountains in Kashmir. Hemp, or early hempcrete, plays a significant role in preserving artwork in the western state of Maharashtra in India (Yadav & Saini, 2022). The use of hemp as a building material is observed in the construction of the bridge in the 6th century in France, however the first modern use of hempcrete is seen in the 1990s in France to renovate work of historic timber-framed buildings (Yadav & Saini, 2022). Due to its carbon-sequestering properly, industrial hemp is a highly successful commercial crop. Researchers believe that it can be used as a cover crop as it uses phytoremediation to remediate contaminated soils and is produced without any pesticides (Ahmed et al., 2022). Furthermore, hemp residue is used as a botanical insecticides, miticides, or inhibitor to soil nematodes and pathogenic fungi (Ahmed et al., 2022). Due to its ability to be resistant to rodents, fungus, and other types of weed, and flourish in a herbicide and pesticide-free environment, more than 30 countries are involved in the global hemp trade. "According to FAO Stat (2018), three major hemp-producing countries by production area are Canada (555,853 ha), North Korea (21,247 ha), and France (12,900 ha) (Ahmed et al., 2022)". The yield of Hemp is dependent on sowing density, harvest time, nitrogen level, moisture in the soil, and can grow up to 0.31 m in a week (Ahmed et al., 2022). Hempcrete is a bio-based material made out of hemp shives, lime, and water (Weber Tradical, 2022). The hemp shives is a by-product of the hemp fiber; it is lightweight and is able to have a very low thermal conductivity due to its high porosity (Barbhuiya & Bhusan Das, 2022). The hemp shives, which is approximately 65-70 percent of the hemp plant, also act as a bio-aggregates, whereas the lime acts as a binder. In a 950 degree celsius limestone kiln, the lime binder is made, which is about 500 degree celsius less than what it takes to make cement. Thus, having a lower embodied energy and a negative embodied carbon requirement in comparison to cement (Barbhuiya & Bhusan Das, 2022). Furthermore, carbon dioxide that is emitted during the carbonation process is absorbed back in the system, resulting in a lower net positive of carbon dioxide emissions (Yadav & Saini, 2022). Hemp production is not only a part of the construction sector, but also belongs to the fiber production industry. "An average production of hemp of approximately 8 t/ha was considered. This stage includes fertilizer production and its transport from the production site to the farm. The emissions due to the fertilizers used in fields are: Ammonia (NH_3) volatilised in the atmosphere (0.02 kg/kg of nitrogen supplied), nitrates (NO_3) emitted in surface water by leaching (40 kg/ha), nitrous oxide (NO_2) emitted in the air (0.0125 kg/kg of nitrogen supplied) and phosphates (PO_4 3) discharged in surface water (0.01 kg/kg of phosphorus supplied)" (Pretot et al., 2014). Therefore, the lack of pesticides and insecticides needed to grow hemp is a great factor to reduce greenhouse gas emissions. Moreover, the lime-based binder is made of 75% of hydrated lime, 15% of hydraulic binder, and 10% of pozzolanic binder. The production of binder involves the extraction of raw materials, transportation of said materials to the transformation site, and mixing and bagging (Pretot et al., 2014). During the transformation stage, the calcination of limestone is converted into carbon oxide (quicklime), which releases carbon dioxide. However, the quicklime is hydrated to turn the oxides into carbon dioxide ($Ca(OH)_2$) (Pretot et al., 2014). According to Pretot et al., during the calcination process, the carbon dioxide released is approximately 594 g per kg of lime (Pretot et al., 2014). The data for the carbon dioxide emissions is taken from the french union of lime producers, which 90% of lime production comes from.

The density of hempcrete is one of its major advantages over concrete. The density is usually dependent on the quality and quantity of the material, shive size, and porosity and compaction energy (Yadav & Saini, 2022). The density is observed between 400 and 500 k/m^3, whereas the spray method is between 200 to 250 kg/m^3. However, if a higher proportion of lime binder is used, a density of 600 to 1000 kg/m^3 is possible. Therefore, hempcrete is stronger and has the capability of good compaction with high density than concrete (Yadav & Saini, 2022). Furthermore, assessments of the life cycle impact of hemp concrete wall were carried out by Boutin et al. and by Ip and Miller, where the life cycle assessment is performed of a sprayed hemp concrete wall (Pretot et al., 2014). "Once the wood-framework is erected, hemp concrete is manufactured by spraying onto plywood formwork. The lime based binder and plant particles are dry-mixed (22 kg of hemp shiv for 44 kg of lime-based binder). The water is added during projection at the nozzle of the spear. This spraying method is well adapted to the construction of buildings that do not exceed approximately 6 m in height. Two types of coating are used: a sand lime coating (for indoor and outdoor) and a hemp-lime coating (for indoor only). The coating is applied manually with a trowel" (Pretot et al., 2014). The study found the lifetime to be over 100 years. Since hempcrete is a non-load bearing material, it is used with a frame that is usually steel, concrete or wood (Pretot et al., 2014). According to Othmura et al. (2002), the density is influenced by the product's spatial orientation in the volume. Since the material composition and the manufacturing processes are manipulated for hempcrete, the density varies greatly, however, higher-density hempcrete is able to hold a higher yield of strength (Barbhuiya & Bhusan Das, 2022). Another great benefit of hempcrete in comparison to concrete is the thermal properties. Since the structure of hemp shives is anisotropic (Barbhuiya & Bhusan Das, 2022). Thermal conductivity increases with moisture and temperature, which is increased by the density of hempcrete. Furthermore, sapropel and other organic binders achieve decreased thermal conductivities. Many other studies indicate that using silica sol binder, instead of lime, increases thermal conductivity to a value of 0.05 W/mK (Barbhuiya & Bhusan Das, 2022). Moreover, the mechanical strength of hempcrete improves as the particle size of hemp shives reduces. Therefore, hempcrete is a structure which further helps and partially acts as an insulation material to bridge gaps between standard insulation panels and structural walls (Barbhuiya & Bhusan Das, 2022). Low thermal diffusivity is necessary for the transmission of heat to slow down, whereas high thermal effusivity increases the amount of stored energy. Thermal conductivity linearly rises with water content, although thermal diffusivity and specific heat capacity work in distinct ways. The influence of formulation, density, and water content on the thermal conductivity of hemp concretes was examined by Collet and Pretot (2014). "Experimental observations and self-consistent scheme modeling are used in the research. At (23 °C; 50% RH), the thermal conductivity of the materials examined ranged from 90 to 160 mW/(m K). The influence of density on thermal conductivity is far greater than that of moisture content. When the density is increased by 67%, the thermal conductivity increases by roughly 54 percent, but it increases by less than 15–20 percent from dry condition to 90 percent RH" (Barbhuiya & Bhusan Das, 2022). Furthermore, hempcrete regulates the indoor temperature due to its vapor permeability, which reduces the load on heating and cooling systems. Thus, increasing insulation capacity, and the quality of the air. The porosity of hempcrete allows the transfer of vapor in the air, which increases the vapor permeability as compared to concrete (Yadav & Saini, 2022). During high humidity, the vapor condenses abc into the liquid state and reverses at times of low humidity, thus acting as a natural humidifier. Furthermore, the waste generated by the hemp power plants is reused in a mix out called mulch (Yadav & Saini, 2022). Therefore, all of hempcrete is recyclable to make new hempcrete with the addition of extra binder since the original lime is carbonated (Yadav & Saini, 2022). Hempcrete is useful to reduce energy usage due to its hygroscopic nature as it stabilizes humidity and temperature and is thermally efficient, decreasing the use of energy (Yadav & Saini, 2022). Furthermore, hempcrete works as a CO_2 sink as the composition of hempcrete is 80% to 90% shiv which absorbs CO_2 in the environment. "Each tonne of hempcrete absorbs about 249 kg of CO_2 over its 100 years of the life cycle" (Yadav & Saini, 2022). However, due to its poor structural performance, it is not a load-bearing construction; a structural wooden frame is necessary for strong support (Yadav & Saini, 2022). Furthermore, hempcrete is used in flooring due to insulating properties, however, the depth of flooring is less than the concrete floors as the main structural layer and subbase are insulating materials, leading to less excavation. Hempcrete in flooring however reduces cost of construction, energy consumption, and is a sustainable environmentally friendly material (Yadav & Saini, 2022). Moreover, hempcrete in roofing construction requires one-inch of space between the top side of hempcrete and the underside of sheeting material.

Plywood, or any other board is used as sheeting material. Roofs made of hempcrete are durable and have a greater tolerance to withstand rain and extreme temperatures without the use of plaster. Thus, reducing energy, cost, and are able to become fire-resistant (Yadav & Saini, 2022).

To simplify, the process of making hempcrete is relatively simple. First, the hemp shiv is separated from the rest of the hemp plant. This is typically done through a process called retting, which involves soaking the plant in water for a period of time. Once the shiv is separated, it is then mixed with lime and water to create the hempcrete mixture. The mixture is then poured into molds or forms and allowed to cure. The curing process can take several weeks to a few months, depending on the conditions. Hempcrete has many potential applications, particularly in the construction industry. It can be used for both load-bearing and non-load-bearing walls, and it can be form into a variety of shapes and sizes. Additionally, it can be used in both new construction and renovation projects. One of the main potential applications of hempcrete is as a replacement for concrete. Concrete is a major contributor to greenhouse gas emissions, as the production of cement (a key ingredient in concrete) is a very energy-intensive process. Additionally, concrete is not a sustainable material, as it requires a large amount of resources to produce. Hempcrete, on the other hand, is a sustainable and low-carbon alternative to concrete. Another potential application of hempcrete is in the insulation industry. The hemp shiv in the mixture is naturally insulating, and when combined with lime, it creates a material that has a very high thermal mass. This means that hempcrete buildings are able to maintain a stable temperature and require less energy for heating and cooling. Overall, hempcrete is a promising building material that has the potential to significantly reduce the carbon footprint of the construction industry. Its low-impact production and its ability to replace traditional materials such as concrete, make it a valuable alternative for the construction industry and a step towards a sustainable future.

References

Ahmed, A. T. M. F., Islam, M. Z., Mahmud, M. S., Sarker, M. E., & Islam, M. R. (2022). Hemp as a potential raw material toward a sustainable world: A review. Heliyon, 8(1), e08753. https://doi.org/10.1016/j.heliyon.2022.e08753

Barbhuiya, S., & Bhusan Das, B. (2022). A comprehensive review on the use of hemp in concrete. Construction and Building Materials, 341, 127857. https://doi.org/10.1016/j.conbuildmat.2022.127857

Belaïd, F. (2022). How does concrete and cement industry transformation contribute to mitigating climate change challenges? Resources, Conservation & Recycling Advances, 15, 200084. https://doi.org/10.1016/j.rcradv.2022.200084

Elfordy, S., Lucas, F., Tancret, F., Scudeller, Y., & Goudet, L. (2008). Mechanical and thermal properties of lime and hemp concrete ("hempcrete") manufactured by a projection process. Construction and Building Materials, 22(10), 2116–2123. https://doi.org/10.1016/j.conbuildmat.2007.07.016

Gourlay, E., Glé, P., Marceau, S., Foy, C., & Moscardelli, S. (2017). Effect of water content on the acoustical and thermal properties of hemp concretes. Construction and Building Materials, 139, 513–523. https://doi.org/10.1016/j.conbuildmat.2016.11.018

Pretot, S., Collet, F., & Garnier, C. (2014). Life cycle assessment of a hemp concrete wall: Impact of thickness and coating. Building and Environment, 72, 223–231. https://doi.org/10.1016/j.buildenv.2013.11.010

What is Hempcrete and How to Use It for Construction ? - Weber Tradical. (2022, May 30). Weber Tradical. https://www.weber-tradical.com/en/hempcrete/what-is-hempcrete-2/

Yadav, M., & Saini, A. (2022). Opportunities & challenges of hempcrete as a building material for construction: An overview. Materials Today: Proceedings, 65, 2021–2028. https://doi.org/10.1016/j.matpr.2022.05.576

Chapter 7: Recycled Plastic
by Chitrini Tandon

Plastic production and then the waste it creates is one of the issues we are currently facing regarding environmental protection. Large quantities of plastic are negatively affecting the environment, and all types of plastic used by humans will eventually become waste and end up in landfills because they cannot be recycled completely or at once. There are roughly 6.5 billion tonnes of plastic waste and discarded rubber produced globally each year (Almeshal et al., 2020). Waste reuse is an important way to recycle and conserve energy in the production process, reduce environmental pollution, lead to a sustainable future, and limit the usage of non-renewable natural resources. Roughly 5% of global anthropogenic carbon dioxide emissions come from the cement industry, making it a critical sector for emission mitigation, and Portland cement production releases carbon dioxide in direct and indirect ways (Clines, 2018). It directly emits carbon dioxide through calcination, which is the process of heating limestone. This process breaks down the calcium carbonate in the limestone into calcium oxide and carbon dioxide. This process leads to about 50% of the carbon dioxide emissions created in cement production. Indirect carbon dioxide emissions come from the fossil fuel that is burned to heat the kiln and account for 40 percent of emissions. The remaining 10 percent comes from the transportation of cement. Plastic waste usage in the material industry has recently become a topic of research, and studies have shown that plastic can be used in concrete. This solution helps to minimize the amount of plastic that ends up in landfills. In recent years, various types of plastics, such as polypropylene (PP), polyethylene terephthalate (PET), and high-density polyethylene (HDPE), have been studied (Allplan, 2018). These studies have shown that unit weight decreases with the addition of plastic, resulting in a reduction in compressive strength, splitting tensile strength, and flexural strength. One recent study done in 2012 found that the recycling rate of plastic is only 8.8 percent, and the rest, which is 91.2 percent, is discarded (Clines, 2018).

One study conducted by Almeshal et al. (2020) examined six different concrete mixtures containing PET as a substitution for sand, with the levels being 0%, 10%, 20%, 30%, 40%, and 50%, to study the effects of this material on the physical and mechanical properties of concrete. The concrete was cast to determine the behavior of fresh and hardened concrete in terms of workability, unit weight, compressive strength, flexural strength, tensile strength, pulse velocity, and fire-resistant behavior. They found that there was a reduction in unit weight and that replacing the sand harmed the concrete's mechanical properties at varying rates. They also supported the idea that plastic waste can be disposed of in specific ratios and can effectively be used in industrial ways. Overall, they found that the workability of the plastic concrete was reduced due to non-uniform and irregularly shaped plastic particles, which negatively affected the workability of the mixes. They also found that the fresh unit weight values for the PET concrete mixtures were 31.6% lower for the PET50 mixture compared to a standard concrete mixture. Additionally, after 28 days of curing, the PET50 mixture had the lowest dry unit weight of all the mixtures tested. They also found that as the proportion of plastic in concrete increased, the compressive strength decreased, with the compressive strength decreasing by 90.6% for the PET50 mixture compared to the reference standard mix. The splitting tensile, flexural strengths, and ultrasonic pulse velocity also decreased. They did find, though, that plastic concrete is recommended for external use and not internal use due to its poor fire resistance. Future research should focus on conducting a feasibility study on the use of plastics as a replacement for sand in concrete production, which should include a comparison between the cost of sand and crushing plastic with the provision of large areas of land due to plastic waste disposal, studies on other sizes and types of recycled plastic, the workability of concrete that contains plastic and the impacts of the addition of admixtures such as super plasticizers, and studies to improve the compressive strength of plastic concrete.

Commonly, the replacement of natural aggregates in concrete with lightweight materials such as plastic has led to a decrease in concrete unit weight, which is a key target in the construction industry. There are many advantages to lightweight concrete, such as having a high thermal insulation response in the buildings and reducing the cost and time spent on handling and manufacturing (Almeshal et al., 2020). Additionally, the impact of an earthquake on a building decreases with decreasing unit weight because earthquake forces are known to be linearly dependent on the self-weight of built structures. Concrete that contains plastic is a good material to use for non-structural elements that do not require high compressive strength, and the high permeability and low absorption of plastic can be used to our advantage in concrete, making it good for applications such as pavements and the floors of sports courts, both of which require good water drainage.

Research is being done all over the world at many universities to understand the best way to mix plastic into concrete. Two students at the Massachusetts Institute of Technology (MIT), Carolyn Schaefer and Michael Ortega, have developed a process to strengthen concrete and plastic waste to reduce CO_2 emissions (Chu, 2017). This is done by using polyethylene terephthalate (PET) bottles obtained from the garbage. They call it "plastic concrete." According to recent estimates, worldwide production of concrete accounts for about five percent of global CO_2 emissions, and more than eight million tonnes of plastic waste flow into the sea. About 45 percent of concrete emissions are caused by the chemical process of converting limestone into cement clinker in cement kilns at 1,500 degrees Celsius. Schaefer and Ortega wanted to find a way to replace this part of cement with finely ground plastic from old plastic bottles. Previous research found that powdered PET weakened the concrete, so the two MIT students wanted to find a way to change the structure of the PET bottles. They achieved this by treating the plastic with harmless gamma radiation. The experiment they conducted showed that the radiation led to chemical rearrangements in the polymer molecules. Interconnections emerge between the molecular chains, making the plastic more rigid and stable and causing porous areas to be filled.

To test the stability of the plastic concrete, the two students mixed cement with a finely powdered version of the irradiated PET bottles and poured the mixture into concrete blocks (Chu, 2017). Through a compression test, they found that the PET-concrete blocks were up to 20 percent more stable than regular cement. This new version of cement can make foundations, bridges, walkways, and barriers more stable and stronger in the future. The addition of 1.5 percent PET powder improves the structure of the concrete.

To make their plastic concrete, the two students first obtained flakes of PET from a local recycling facility; they then sorted through the flakes to remove pieces of metal and debris (Chu, 2017). Schaefer and Ortega brought the pieces of plastic to the basement of MIT's Building 8, which has a cobalt-60 irradiator that emits gamma rays. Gamma rays are typically used commercially to decontaminate food, and there is no residual radioactivity from this type of radiation. The research team then exposed batches of the flakes to low or high levels of gamma rays, ground each batch of flakes into a powder, and mixed the powders with a series of cement paste samples. The cement was a traditional Portland cement powder and one of two common mineral additives, either fly ash (a byproduct of coal combustion) or silica fume (a byproduct of silicon production). The samples contained about 1.5% irradiated plastic. After the samples were mixed with water, they poured the mixtures into cylindrical molds and allowed them to cure. They then removed the molds and subjected the resulting concrete cylinders to compression tests. This allowed them to measure the strength of each sample and compare it with similar samples made with regular, non irradiated plastic and samples with no plastic. Their results showed that in general, samples with regular plastic were weaker than those without plastic, and the concrete with fly ash or silica fume was stronger than the concrete made with only Portland cement. They also found that the cement with plastic and fly ash was stronger than regular concrete, and it increased the strength by up to 15%. This was particularly true in samples with high levels of irradiated plastic.

The research team then used multiple imaging techniques to examine why the irradiated plastic created a stronger version of concrete (Chu, 2017). They took their samples to Argonne National Laboratory and the Center for Materials Science and Engineering (CMSE) at MIT and analyzed them using X-ray diffraction, backscattered electron microscopy, and X-ray microtomography. From their images, they found that the samples that contained irradiated plastic showed crystalline structures with more cross-linking, or molecular connections. The images also showed that the crystalline structures blocked pores in the concrete, making it denser and therefore stronger.

There are two more advantages to using plastic concrete. The first is that replacing a part of the cement with plastic led to fewer emissions from concrete production. Additionally, it led to fewer PET bottles that ended up in landfills and garbage (Allplan, 2018). Michael Short from the Department of Nuclear Science and Engineering states, "The technology helps to make buildings more stable while at the same time reducing the quantity of waste and CO2 emissions." Kunal Kupwade-Patil, a research scientist in the Department of Civil and Environmental Engineering who was part of the project, says, "At a nano-level, this irradiated plastic affects the crystallinity of concrete" (Chu, 2017). "The irradiated plastic has some reactivity, and when it mixes with Portland cement and fly ash, all three together give the magic formula, and you get stronger concrete." She also states, "We have observed that, within the parameters of our test program, the higher the irradiated dose, the higher the strength of concrete, so further research is needed to tailor the mixture and optimize the process with irradiation for the most effective results." "The method has the potential to achieve sustainable solutions with improved performance for both structural and nonstructural applications." The team plans to continue to experiment with different types of plastics, use different doses of gamma radiation, and determine their effects on concrete. Additionally, plastic aggregates have five times lower thermal conductivity compared to silica-based aggregates, which can help control heat loss during the summer months in buildings and heat gain in the winter (Clines, 2018).

Another study by researchers at MSU's Norm Abjornson College of Engineering also found that plastics treated with specific bacteria can be added to concrete in large amounts without affecting the material's strength (Montana State University, 2021). Typically, adding plastic or other filler material can disrupt the mix of sand and reduce concrete's ability to bind together and support heavy loads. The MSU team has found that using bacteria to coat the plastic with a thin mineral layer allows it to bond better with cement. In their study, they found that concrete samples with up to 5 percent of the bacteria-treated plastic had roughly the same strength as traditional concrete. The experiment was done in MSU's Center for Biofilm Engineering, where the researchers immersed the plastic in a water-based solution that contained the harmless bacteria Sporosarcina pasteurii. This bacteria grows on the surface and forms a biofilm. The microbes that are left in the solution for 24 to 48 hours consume the added calcium and urea, which gives the plastic a thin, white coating of calcite. This is a hard mineral made of limestone. The plastic is then mixed into small concrete cylinders and crushed with specialized equipment to measure the strength. The next step is to determine the material's long-term durability and how the process can be scaled up to be manufactured in usable quantities.

One other study focuses on exploring the effectiveness of gamma-irradiated plastics in cement paste. The irradiated plastics are paired with mineral additives, and the aim of the study is to find the best combination for producing the strongest concrete. This is done through an internal microstructure analysis. One potential way of restoring some of the lost strength in concrete is to use irradiation (Schaefer et al., 2018). PET is a semi-crystalline polyester that has an isotropic microstructure due to its glassy amorphous composition, which makes it one of the most studied polymers. Irradiation of PET leads to chain scission and crosslinking. Chain scission increases the degree of crystallinity in PET with the gamma radiation dose, which decreases the molecular weight. Having a lower molecular weight leads to better molecular mobility, which facilitates the ordered arrangement of molecules in crystalline structures. PET with a higher crystallinity has a higher modulus, toughness, stiffness, strength, and hardness. Additionally, radiation can induce crosslinking, which can strengthen the chemical structure of the compound. Both scission and crosslinking can occur at the same time, and both can improve strength. In this study, a series of cement paste samples were prepared using Portland cement, fly ash, and silica fume. They obtained plastic flakes from a recycling facility that was manually sorted. Half the sample was irradiated at a low dose and the other half at a high dose, after which the samples were further crushed. The three plastic types were: no dose, low dose, and high dose. The cement, once prepared, went through a series of tests. The tests were a compression test, which provides immediate information on the bulk strength of the specimen; X-RAT diffraction, in which specimens were crushed into a powder to allow the crystalline phases to be identified and quantified; backscattered electron (BSE); and energy dispersive spectroscopy (EDS) analysis. BSE was performed on polished samples and was used to examine the effects of the additives on the structure and chemistry of the cement. EDS was used for chemical characterizations and can measure the analytical capabilities from measurements of the photon energy emitted from the cement. The last test done was X-ray microtomography, in which fragments of the hardened cement were examined using X-ray microtomography, which creates a three-dimensional image.

The researchers found through the compression test that cement with a high dose of irradiated plastic additive showed improved strength compared to the other two groups (Schaefer et al., 2018). They suggest that a potential solution for giving cement some of its strength back when plastic is added is through gamma irradiation, and it shows that irradiation of plastic can help partially recover some of the strength of cement. Additionally, the results from the XRD analysis showed that the differences in C-S-H and C-A-S-H phase formation from the addition of both irradiated plastic and mineral additives helped to form high-density phases, which also aided in higher strength levels. It was seen that there was a presence of the gismondine phase, which is potentially the reason that irradiated plastic potentially changed the structure of C-S-H and C-A-S-H gels. Moreover, the BSE analysis showed that there was a higher level of alumni in the fly ash samples, which could potentially help form the high-density phases that led to higher relative strengths in the samples. The X-ray microtomography showed that samples with a high dose of irradiated plastic led to a decrease in segmented porosity, which supports the idea that irradiated plastic can act like a pore-blocking agent. This effect is in addition to the formation of chemical phases such as CSH and CASH, which contribute to the densification of the cement. This approach supports the idea that irradiated plastic can be used in cement and that it can be incorporated into developing new codes and standards for cementitious materials.

Another study done at the University of Bath and Goa Engineering College in India has also found that replacing 10 percent of the sand in concrete with plastic can help reduce the plastic waste on the streets of India and is one way to deal with India's shortage of sand (Clines, 2018). The researchers tested five types of plastic particles and found that when sand was replaced with plastic bottles that were of similar size and shape, it resulted in concrete that was almost as strong as regular concrete. This discovery could potentially save 820 million tonnes of sand per year as well as reduce plastic waste levels. John Orr, who is the principal investigator and a lecturer at Cambridge University, states, "Typically, when you put an inert, man-made material like plastic into concrete, you lose a bit of strength because the plastic material doesn't bond to the cement paste in the material in the same way that a sand particle would." The key challenge here was to strike a balance between achieving a small reduction in strengths and using an appropriate amount of plastic to make it worthwhile."It is really a viable material for use in some areas of construction that might help us tackle issues of not being able to recycle the plastic and meeting a demand for sand." Additionally, Richard Ball, who is a co-researcher in the University of Bath's Department of Architecture and Civil Engineering, has also stated that "even when the reduction in performance prohibits structural applications, the lower-tech uses such as paving slabs may be viable." With upwards of 40% of plastic ending up in landfills in India, this new discovery can greatly aid the plastic waste issue (Cousins, 2019).

A company called ByFusion Global has created what they call ByBlocks (Harvey, 2022). These concrete blocks are made from hard-to-recycle and recyclable plastics, such as plastic bags, yogurt containers, and water bottles. Unlike regular concrete blocks, these blocks do not break, crumble, or crack. According to the ByFusion Global website, "ByBlock is the first construction-grade building material made entirely from recycled (and often non-recyclable) plastic waste." These concrete blocks can be installed without the use of glue or adhesives. The blocks are made from a steam-based process called ByFusion, which does not use any chemicals, additives, or fillers. ByBlocks can be used to swap out concrete blocks in retaining walls, foundations, inner walls, terraces, and more.

It is estimated that more than 20 billion tonnes of concrete are produced globally each year; this makes it the second most consumed substance in the world after freshwater (Cousins, 2019). There are many environmental concerns related to the process of creating concrete and the plastic waste that is produced globally. Using plastic waste in cement is killing two birds with one stone, and many researchers around the world have realized that. While using plastic in concrete is not a popular practice right now, there is a lot of research happening around the world, and one day it might be used frequently.

References

Allplan. (2018). Plastic as a replacement for cement: Stronger eco-cement in the future? ALLPLAN Blog - News zu BIM & CAD. Retrieved January 12, 2023, from https://blog.allplan.com/en/eco-cement#:~:text=Used%20plastic%20bottles%20could%20make,result%20from%20the%20concrete%20production.

Almeshal, I., Tayeh, B. A., Alyousef, R., Alabduljabbar, H., & Mohamed, A. M. (2020). Eco-friendly concrete containing recycled plastic as partial replacement for sand. Journal of Materials Research and Technology, 9(3), 4631–4643. https://doi.org/10.1016/j.jmrt.2020.02.090

Chu , J. (2017). MIT students fortify concrete by adding recycled plastic. MIT News | Massachusetts Institute of Technology. Retrieved January 12, 2023, from https://news.mit.edu/2017/fortify-concrete-adding-recycled-plastic-1025

Clines, K. (2018, October 14). How replacing sand with plastic in concrete solves problems. Equipment World. Retrieved January 31, 2023, from https://www.equipmentworld.com/better-roads/article/14970100/how-replacing-sand-with-plastic-in-concrete-solves-problems

Cousins, S. (2019). Sustainability double puts plastic in concrete. RIBAJ. Retrieved January 31, 2023, from https://www.ribaj.com/products/extreme-spec-waste-plastic-in-concrete-university-of-bath-goa-engineering-college

Harvey, O. (2022, February 18). Hard-to-recycle plastics are now being made into Zero waste "concrete" blocks. Apartment Therapy. Retrieved January 31, 2023, from https://www.apartmenttherapy.com/recycled-plastic-concrete-37041922

Montana State University. (2021, May 26). Researchers find potential use for recycled plastic in concrete. Phys.org. Retrieved January 22, 2023, from https://phys.org/news/2021-05-potential-recycled-plastic-concrete.html

Schaefer, C. E., Kupwade-Patil, K., Ortega, M., Soriano, C., Büyüköztürk, O., White, A. E., & Short, M. P. (2018). Irradiated recycled plastic as a concrete additive for improved chemo-mechanical properties and lower carbon footprint. Waste Management, 71, 426–439. https://doi.org/10.1016/j.wasman.2017.09.033

Chapter 8: Mycelium
by Maggie Wang

Introduction

Concrete, one of the most popular composite materials across the world, comprises human-made structures ranging from skyscrapers to the road beneath our feet. Despite concrete's ubiquitous nature, alongside the fact that the material is both cheap and durable, it is a massive contributor to climate change globally. As mentioned in previous chapters, construction and associated materials account for 40% of the world's annual carbon dioxide emissions, with concrete being responsible for a whopping 8% of total emissions (Avery, 2021). These shocking statistics highlight the urgent need for research into sustainable and carbon-neutral alternatives to concrete. Especially with the spike in urbanization and populations moving towards more urban areas, this presents an increased demand for more habitat construction (Modanloo et al., 2021).

In order to reduce the impact that urbanization and other forms of construction have on climate change, many newer studies have been focused on finding sustainable alternatives within microorganisms such as bacteria and fungi. In the recent past, bacteria has been utilized as a means of sealing fractures within various oil and gas wells that would otherwise be difficult to reach (Yellowstone Public Radio, 2021). Not only is this beneficial to recollect oil and gas to ensure that none is wasted, but also because methane released in the environment can be highly detrimental to the health of the Earth's greenhouse layer (Yellowstone Public Radio, 2021). In other cases, a particular interest has been taken in mycelium, the lower part of vegetative fungi that can be used to develop various mycelium-based composite materials. Given that these materials are both sustainable and biodegradable, the following chapter will take a deeper dive into exploring what mycelium is, its various properties, how it grows, how its growth can be tailored for different construction purposes, benefits and consequences of using mycelium-related materials, and different architectural endeavors that have real-world applications of mycelium building.

An Introduction to Mycelium: The Ubiquitous Fungi

Mycelium is the main body (and vegetative portion) of fungi, often branching extensively to encompass large volumes of space. In fact, mycelium is the largest living organism on Earth, with the largest being that of the mycelium network in the Oregon Blue Mountain range (Hawksworth, 2001). Mycelium is primarily composed of hyphae, which are the filamentous portion (Haneef et al., 2017). These hyphae consist of elongated cells that are separated by septa (porous walls). These hyphae are enclosed by a cell wall that is very physiologically important for the mycelium (Haneef et al., 2017). Not only does the cell wall protect the hyphae from external stressors, but it also contributes to a large portion of the characteristic mechanical strength of mycelium (Haneef et al., 2017). This cell wall is composed of various proteins, glucans, and most importantly, chitin (Haneef et al., 2017). Chitin, a large and widely branched polysaccharide, provides mycelium with many of its physical characteristics (Haneef et al., 2017). This overall composition makes mycelium a fibrous composite material that can be utilized for building purposes.

All fungi belong to the class of heterotrophic organisms, meaning that they cannot produce their own foods (Modanloo et al., 2021). As a result, they require organic compounds for sustenance and growth, thus forming a symbiotic relationship with these substrate compounds (Modanloo et al., 2021). Since the mycelia are the body and root of the fungi, they are able to intake and absorb nutrients from the nearby environment (Modanloo et al., 2021). The hyphae in the mycelium release enzymes to break down these nutrients into monomers (small units) for absorption (Modanloo et al., 2021). This process and consequent binding between the mycelium and its food substrates creates a material known as a mycelium-based composite. In this way, they are able to form a network of fibrous mycelium branches that continue to grow and propagate through further absorption.

Interfering with the Mycelial Growth Process

Given that mycelium has such a unique structure and physical composition, it makes it a great candidate to further explore how these can be utilized in materials construction. Interestingly, there are a multitude of factors within the mycelial growth process that can be altered to obtain different physical properties. These factors can thus be tailored depending on the type of structure required. Firstly, by altering the type of fungi that the mycelium is cultivated and extracted from, the chemical and biological composition of the final mycelium-based composite is altered (Modanloo et al., 2021). Secondly, the environmental conditions (i.e., nutrients available, temperature, the humidity of the environment, pH, and aeration) can greatly impact the final product as well (Modanloo et al., 2021). Finally, the formation and storage techniques of the mycelium can also be altered. As these techniques occur after the processing of the mycelium, there is great diversity in what can be tried. For example, these composite materials can be hot-pressed or cold-pressed to alter the density and porosity values (Modanloo et al., 2021).

A study by Haneef et al. has focused on controlling an environmental condition — the substrates that are available to the mycelium — and examining the effect that this has on the physical composition of the mycelium (2017). The fungi that the mycelium was later harvested from were the Ganoderma lucidum and Pleurotus osteratus species. Given that these species are known for containing enzymes that are able to break down plant components that normally may be harder to hydrolyze, they were the optimal choice for the study. These fungi were fed 2 different bio-substrates, one being cellulose alone, and the other being cellulose/potato-dextrose. Prior to commencing the study, the authors noted that the cellulose/potato-dextrose substrate is known to be more easily digestible by mycelium, as there is a higher quantity of simple sugars in the substrate. As predicted, the substrate did have substantial impacts on the physical composition of the final mycelium structures, such as polysaccharide, lipid, protein, and chitin concentrations (Haneef et al., 2017). The mycelium that absorbed cellulose was found to contain higher chitin concentrations and lower levels of elongation (Haneef et al., 2017). On the other hand, cellulose/potato-dextrose absorption led to lower chitin concentrations and more elongation (Haneef et al., 2017). As a result, the authors concluded that when the substrates are less digestible by the mycelium, the final material tends to be stiffer and less ductile (Haneef et al., 2017). Regardless of the substrate used, all of the samples were found to be highly hydrophobic, thermally stable (high degradation temperatures), have low water uptake levels, and require minimal energy levels to produce (Haneef et al., 2017). Overall, this study has demonstrated that mycelial growing patterns can be altered to tailor the final product required, alongside providing that mycelium-based composite materials are a fantastic sustainable alternative that paves the way for a promising avenue of research.

Another study by Modanloo et al. has tackled the same concept but at a different level of the growth process — the post-processing step (2021). The authors focused on investigating various means of digital design and fabrication to tailor mycelium-based composite materials for architectural purposes, using an additive technique to form the structure of the final material (Modanloo et al., 2021). This technique is important, as the growth of mycelium is dependent on the presence of air. In order to create a complex desired final shape, the authors had to form the mycelium-based material into a desired mold and then allow the outside to access air (Modanloo et al., 2021). The authors first started by conducting a systematic study on prior components of the mycelial growth process, altering the effects of substrate on the mechanical strength of mycelium-based composite blocks (Modanloo et al., 2021). After testing multiple substrates, they utilized a final mixture of shredded paper, oyster mushroom, wheat bran, and guar gum, which they found to have adequate mechanical strength (Modanloo et al., 2021). Later, they used a digital design process to create a tilted arch as a means of testing the material created as well as its strength (Modanloo et al., 2021). This study has major implications for further architectural work with mycelium-based composite materials, providing the scientific community with solid foundational work on various design and computational mechanisms. In addition, these materials are lightweight and regenerative, which helps to address various shortcomings that some architectural structures currently possess, not to mention compostable and sustainable. However, the authors noted that the future may prevent additional challenges with scaling up the level of this construction (Modanloo et al., 2021).

Building with Mycelium-Based Composite Materials

Current researchers at Montana State University have been investigating how to utilize mycelium-based composite materials as an alternative to cement. In order to do this, the team begins by working with a large syringe system that is filled with different types of sand (Yellowstone Public Radio, 2021). The researchers then attach tubes to the syringes and use these to pump a special substance into the syringe, which contains the fungus and a liquid that suspends the fungus (Yellowstone Public Radio, 2021). The mycelium continues to branch within the syringe, propagating into a large network. Once this happens, the cementing process begins, where the researchers add nutrients, calcium, and bacteria into the syringe (Yellowstone Public Radio, 2021). Upon this addition, calcium carbonate forms, which serves as the main component that replaces cement (Yellowstone Public Radio, 2021). This cement alternative is a focal point of interest in the sustainability community because it is carbon neutral and biodegradable. In this way, the calcium carbonate can be decomposed and then used as cement or concrete for other building purposes, serving essentially as a 'recyclable cement'. Furthermore, this cement alternative is also particularly desirable because the entire process can be done at room temperature, which rivals the high temperatures and energy inputs that are often required with typical cement and concrete production (Yellowstone Public Radio, 2021). However, despite this new alternative being a promising avenue of research, it also presents unanswered questions that revolve around the durability of this 'cement' as well as potential adverse reactions that the 'cement' may have with other environmental materials (i.e., metal, plastic, etc.) (Yellowstone Public Radio, 2021).

Another group at The Living firm looks into utilizing mycelium-based composite materials to make "mushroom bricks" (Avery, 2021), To start off this process, a mixture of mycelium and agricultural waste (e.g. straw or corn husks) are grown together (Avery, 2021). After approximately two weeks, the mixture is at a point where the agricultural husk has been fully colonized (Avery, 2021). At this point, the fungus in the mixture must be killed, either through heat or chemical treatment (Avery, 2021). Once these bricks have been formed, they can be stacked and used for architectural and construction purposes. Beyond the benefits previously stated regarding the sustainable and biodegradable nature of mycelium-based composite materials, these mushroom bricks also present additional benefits as they are potentially self-repairing. Another researcher, Phil Ayres, at Copenhagen's Center for Information Technology and Architecture, has noted that the mycelium does not necessarily need to be killed at the end of the production process (Avery, 2021). Rather, the mycelium could be allowed to continue to propagate and grow into the desired shape, rather than consistently making brick forms (Avery, 2021). In addition to this, mycelium could be self-repairing when there is damage to the architectural structure (e.g. hole in a wall), as the mycelia could regrow in those areas if kept alive (Avery, 2021). With this suggestion though, comes the possibility of mycelium overgrowth that may potentially compromise the integrity of the architectural structure. In order to combat this, Ayres has suggested that mycelium-based walls be made in a sandwich fashion; that is, two layers of dead mycelium on the outside and one layer of live mycelium on the inside (2021). The inner layer would remain dormant and prevented from overgrowing due to a lack of access to water, however it could be activated when needed (i.e., when reparative changes are necessary) (Avery, 2021). Interestingly, mycelium also responds to electrical impulses, which paves the way for fascinating applications in 'smart' home systems, such as automatic windows and doors (Avery, 2021).

The Present and the Future of Mycelium in Building

Given that fungus-based materials such as mycelium have become increasingly popular in architectural and construction industries due to its sustainable and biodegradable nature, the following section will take a deeper dive into current and future projects that utilize mycelium-based composite materials. Current research is not only investigating the effectiveness of these materials when put into practice, but also innovative and creative ways in which these materials can be applied in design. This research is becoming increasingly important considering the shocking statistics regarding CO2 emissions within the building sector: this industry is responsible for 38% of CO2 emissions (Almpani-Lekka et al., 2021).

Current mycelium-based architecture has been largely founded by companies such as 'Ecovative', which focus on substituting environmentally damaging products (plastic materials, insulation, etc.) with carbon negative products (Almpani-Lekka et al., 2021). Since the rapid development of these companies, fungal architecture has been progressing quickly, building off of prior designs that other innovators have created.

Firstly, a notable structure is the Hy-Fi pavilion that was built in the MoMa art museum in 2014 by the company Ecovative (Almpani-Lekka et al., 2021). By utilizing numerous mycelium-based blocks, the team created a design that looks like a group of vertically clustered cylinders. Given that this project used 10,000 blocks, it serves as the largest mycelium-based project built to this day (Almpani-Lekka et al., 2021). It incorporates many aspects of ergonomic and innovative design, such as ensuring that adequate ventilation is created by spacing the bricks apart, utilizing light refractive film, and using scaffolds to ensure that the building maintains its structural integrity (Almpani-Lekka et al., 2021). Furthermore, this structure attests to the biodegradable nature of mycelium-based composite materials, as after the exhibition ended, the structure was shredded and degraded.

Secondly, a similar pavilion structure was created for a contemporary art exhibition in Southwest India called the Kochi-Muziris Biennale (Almpani-Lekka et al., 2021). In particular, Studio Beetles 3.3 and Yassin Arredia Design collaborated to create this. This structure was a living structure, where the mycelium grew over cavities in the structure and then dried through exposure to sunlight (Almpani-Lekka et al., 2021). Not only was the structure lightweight, but it could actually be dismantled and reassembled when it was time for other temporary events.

Thirdly, a structure called the MycoTree was a collaboration between various construction and research groups (Sustainable Construction KIT, Block Research Group, etc.) (Almpani-Lekka et al., 2021). This structure was installed for the Seoul Biennale for Architecture and Urbanism. The design was composed of mycelium-based materials that branched towards the ceiling, utilizing and optimizing parametric design (Almpani-Lekka et al., 2021). The complicated structure was created using mycelium materials that were shaped within particular molds.

Fourthly, El Monolito Micelio was a construction experiment designed and built by the Georgia Institute of Technology School of Architecture (Almpani-Lekka et al., 2021). This was another pavilion-like structure that focused on casting mycelium-based composite materials on the top of the pavilion. Through use of computational design, the architects designed this structure to absorb compressive forces (Almpani-Lekka et al., 2021). Despite being an innovative and unique structure in the world of mycelium-based building, the structure was left to dry after being casted, resulting in superficial cracks forming in different parts of the pavilion (Almpani-Lekka et al., 2021).

Fifthly, the MY-CO Space was designed by the ArtSci collective MY-CO-X, a collaboration between Vera Meyer and Sven Pfeiffer, a biotechnologist and architect, respectively (Almpani-Lekka et al., 2021). This project was an exhibit that comprised a living and learning space for residents to test out, as it was part of a larger German exhibition that revolved around creating habitable sculptures. Interestingly, the structure was composed of hundreds of small mycelium-based elements attached to a plywood substructure and base (Almpani-Lekka et al., 2021).

Beyond creating artistic structures, mycelium-based composite materials have also been thought to be useful for other practical reasons as well. Given that space travel is often hindered by astronomical energy costs, the concept of bringing heavy construction materials to the Moon or Mars to build habitats is often incomprehensible. However, since mycelium-based materials are lightweight, there have been suggestions to build mycelia habitats instead (Avery, 2021). In addition to this suggestion, some companies are also optimizing the use of mushrooms for everyday products and uses, such as insulation and even clothing. IKEA and Dell are utilizing mushroom-based packaging material as a carbon neutral alternative to plastic (Avery, 2021). A Dutch company has initiated a project called The Living Cocoon, where toxic materials can be safely buried and the time required for the decomposition process is shortened (Avery, 2021). In fact, this structure can even aid in the process of growing new trees. Finally, a particularly interesting and innovative case of utilizing mushrooms in building, is by Katy Ayers, a Nebraska resident that created a canoe made out of mushrooms (Avery, 2021). Due to mycelium's buoyant and waterproof nature, Ayers was able to effectively create a two-seater boat that grew in just two weeks. In collaboration with Nebraska Mushroom, they created a skeleton that served as the substructure for the mycelium-based composite material, and then allowed the mycelium mold to grow in a shell-like structure (Avery, 2021). These works of architecture all speak to the ubiquitous nature of mushrooms, which could potentially serve as the future of carbon neutral, biodegradable, and sustainable construction.

Conclusion

As aforementioned, there is an integral need in the construction industry at present, to increase focus and research on environmentally friendly and sustainable materials for building. Mycelium is a particularly promising building composite material, given that it has numerous benefits in being lightweight, self-sustaining and growing, biodegradable, and plentiful in our ecosystem. Furthermore, it is an energy-efficient material, requiring very little energy for construction purposes and creating negligible carbon emissions. As a result, it is key to continue investigating these materials and their use in architecture and various other useful endeavors. This chapter discussed the biological mechanisms of mycelium growth, the structure of mycelia, current research in tailoring mycelium-based construction techniques, various benefits and downfalls of mycelium-construction, and architectural projects that have primarily used fungi. As seen from these creative structures, this is a very promising avenue of research that could quite possibly be the future of carbon neutral construction. In fact, with more research and innovation, it is possible that we could all live in mushroom-based houses in the future.

References

Almpani-Lekka, D., Pfeiffer, S., Schmidts, C., & Seo, S. (2021). A review on architecture withnfungal biomaterials: The desired and the feasible. Fungal Biology and Biotechnology, 8,17. https://doi.org/10.1186/s40694-021-00124-5

Avery, D. (2021, February 3). Bricks made from MUSHROOMS could soon replace cement. Mail Online. https://www.dailymail.co.uk/sciencetech/article-9220581/Bricks-MUSHROOMS-soon-replace-cement-self-repair.html

Hawksworth, D. L. (2001). The magnitude of fungal diversity: The 1.5 million species estimate revisited* *Paper presented at the Asian Mycological Congress 2000 (AMC 2000), incorporating the 2nd Asia-Pacific Mycological Congress on Biodiversity and Biotechnology, and held at the University of Hong Kong on 9-13 July 2000. Mycological Research, 105(12), 1422–1432. https://doi.org/10.1017/S0953756201004725

Haneef, M., Ceseracciu, L., Canale, C., Bayer, I. S., Heredia-Guerrero, J. A., & Athanassiou, A. (2017). Advanced Materials From Fungal Mycelium: Fabrication and Tuning of Physical Properties. Scientific Reports, 7, 41292. https://doi.org/10.1038/srep41292

Modanloo, B., Ghazvinian, A., Matini, M., & Andaroodi, E. (2021). Tilted Arch; Implementation of Additive Manufacturing and Bio-Welding of Mycelium-Based Composites. Biomimetics, 6(4), 68. https://doi.org/10.3390/biomimetics6040068

Yellowstone Public Radio. (2021, February 25). MSU Researchers Exploring Cement Alternative Using Fungi, Bacteria. YPR. https://www.ypradio.org/environment-science/2021-02-24/msu-researchers-exploring-cement-alternative-using-fungi-bacteria

Chapter 9: Bamboo
by Kanish Baskaran

Bamboo is a perennial plant native to Asia, Africa, and South America. It has many applications in building, woodwork, textiles, and other industries and grows rapidly. Bamboo is a sustainable resource that may be grown without pesticides or fertilizers (Dillon, n.d.).

Bamboo can grow up to 91 centimeters every day, ranking it one of the world's fastest-growing plants (Hebel, 2018). It reaches maturity in three to five years, but many hardwood trees require decades to attain maturity (Schroder, n.d.). This makes bamboo a renewable and sustainable resource that can be grown rapidly and without the use of toxic chemicals (Hebel, 2018). Bamboo is used as an alternative for conventional building materials such as wood and concrete in construction. It is sturdy and long-lasting, and can be used for a range of applications, including flooring, paneling, and scaffolding (Hebel, 2018). Bamboo is also used in the furniture and textile industries to create garments and other materials. Bamboo is also a popular landscaping material, as it can be utilized to construct privacy barriers, shrubs, and even living fences (Hebel, 2018). It is also used as an attractive plant in gardening due to its unusual appearance and rapid growth. In addition to its practical applications, bamboo has numerous environmental advantages. It is a carbon sink, meaning it collects co2 from the atmosphere and can help mitigate climate change (Hebel, 2018). In addition to enhancing soil quality and preventing erosion, bamboo is an important plant for conservation efforts (Hebel, 2018). This chapter examines bamboo and its numerous construction applications.

Structural Composition and Harvesting of Bamboo

Bamboo is a grass-related perennial woody plant (Poaceae) (Janssen (auth.), 1991). It consists of an underground network of rhizomes and culms (stems) that sprout from these rhizomes. The hollow culms of bamboo provide to the plant its characteristic light weight and durability. The culms of bamboo may grow up to 91 cm every day, making it one of the world's fastest-growing plants. It reaches maturity in three to five years, but many hardwood trees require decades to attain maturity (Janssen (auth.), 1991).

The structure of bamboo consists of three components: rhizomes, culms, and leaves. The rhizomes are horizontally growing underground stems that produce new shoots (Janssen (auth.), 1991). The culms are the hollow, segmented stems that grow above ground. The leaves are long and narrow, and they emerge from the culm nodes (Janssen (auth.), 1991).

Bamboo is an adaptable and resilient plant that can thrive in a variety of climates and soil types. It may be cultivated in tropical, subtropical, and temperate settings and can thrive in a variety of soil types, including clay, sand, and rocky terrain (Janssen (auth.), 1991). Bamboo is a resilient plant that can live in regions with heavy precipitation, strong winds, and high temperatures (Janssen (auth.), 1991). Additionally, bamboo is a comparatively low-maintenance crop. It thrives in regions where other crops fail with minimal fertilization. Additionally, bamboo may be cultivated without the use of pesticides or fertilizers, making it an environmentally preferable alternative to other crops (Janssen (auth.), 1991).

There are multiple methods for cultivating bamboo, including planting from seed, planting from rhizomes, and planting from cuttings (Schroder, n.d.). Planting from rhizomes is the most prevalent approach since it is faster and more efficient than planting from seeds (Neely, 2022). When soil moisture is high during the rainy season, rhizomes should be planted. Similarly to rhizomes, cuttings can also be obtained from mature plants and planted. Bamboo can also be propagated by tissue culture, which involves cultivating tiny pieces of plant tissue in a sterile environment, such as leaves or stems (Neely, 2022). This method is more labor-intensive and costly than conventional propagation techniques, yet it allows for the mass production of high-quality plants (Neely, 2022).

Bamboo is a flexible and ecological plant that has numerous uses in the building, textile, and other sectors (Neely, 2022). It consists of an underground network of rhizomes and culms (stems) that have sprouted from these rhizomes. Bamboo is a resilient plant that can live in regions with heavy rainfall, strong winds, and high temperatures (Neely, 2022). Bamboo may be cultivated in a variety of temperatures and soils and requires little care. There are multiple methods for cultivating bamboo, including planting from seed, planting from rhizomes, and planting from cuttings. With correct cultivation and care, bamboo can produce high-quality plants that can be used for human and environmental benefit (Neely, 2022).

Problems with Concrete

Concrete is a widely used building material that has proven to be effective in the construction of various structures, including buildings, bridges, and roads (Encyclopedia Britannica, 2022). However, despite its widespread use, there are a number of issues associated with concrete as a building material (Encyclopedia Britannica, 2022).

Its substantial carbon footprint is one of the most significant issues with concrete. The production of concrete is energy-intensive and a significant contributor to global warming and climate change due to the emission of greenhouse gasses (Encyclopedia Britannica, 2022). In addition, the production of concrete necessitates large quantities of cement, sand, gravel, and water, which can have a significant impact on local ecosystems and natural resources (Encyclopedia Britannica, 2022).

Additionally, concrete is susceptible to cracking and deterioration over time. Concrete is prone to cracking due to temperature changes, moisture, and other environmental factors (Encyclopedia Britannica,

2022). This can result in structural damage and compromise the structure's safety and stability. In addition, concrete is also susceptible to degradation due to chemical reactions, such as corrosion and alkali-silica reaction, which can lead to cracking and spalling of the concrete surface (Encyclopedia Britannica, 2022).

The weight of concrete is another problem that can be a challenge in construction. Concrete is heavy, which can make it difficult to transport and handle (Encyclopedia Britannica, 2022). This can be especially difficult in areas with limited access or when the structure must be constructed on soft or unstable ground. In addition, the weight of concrete can also be a challenge in seismic areas, as it can make the structure more susceptible to damage in the event of an earthquake (Encyclopedia Britannica, 2022).

In addition to these issues, concrete has poor insulating properties, which can be problematic in cold climates (Encyclopedia Britannica, 2022). This can lead to increased heating and cooling costs, as well as increased energy consumption. In addition, concrete is a relatively poor conductor of heat and electricity, making it difficult to install plumbing and electrical systems within concrete structures (Encyclopedia Britannica, 2022).

In conclusion, despite its widespread use, concrete has a number of disadvantages as a construction material. From its high carbon footprint and susceptibility to cracking and degradation, to its weight and limited insulation properties, these problems can be significant challenges in construction. As a result, there is a growing demand for alternative building materials that are more durable, sustainable, and affordable (Encyclopedia Britannica, 2022).

Use of Bamboo as a Concrete Alternative in Construction

Bamboo is a flexible and sustainable plant with numerous construction applications. Its strength, resilience, and rapid development make it a suitable building material for a range of purposes (Nurdiah, 2016). In addition to being a renewable and eco-friendly resource, bamboo can be produced rapidly and without the use of dangerous chemicals (Nurdiah, 2016).

As a substitute for conventional building materials such as wood and concrete, bamboo is one of its primary applications in the construction industry (Nurdiah, 2016). Bamboo is robust and long-lasting, and it may be used for a variety of purposes, including flooring, paneling, and scaffolding. In addition to being used to create houses, bridges, and other buildings, bamboo is also employed in the building industry (Nurdiah, 2016). Bamboo is also utilized in the creation of eco-friendly and sustainable structures (Nurdiah, 2016). Bamboo is a carbon sink, meaning it collects carbon dioxide from the atmosphere and contributes to mitigating the effects of climate change. In addition to enhancing soil quality and preventing erosion, bamboo is an important plant for conservation efforts (Nurdiah, 2016).

Additionally, bamboo can be utilized to make green roofs and living walls (Dillon, n.d.). On the top of a building, bamboo plants can be grown to offer insulation, minimize the heat island effect, and enhance air quality (Dillon, n.d.). Bamboo can also be grown on the outer walls of a structure to form living walls, which can enhance the building's appearance and provide additional insulation (Dillon, n.d.). Geodesic domes, which are structures composed of a network of triangles, can also be constructed from bamboo. These structures are renowned for their stability and strength, and they can be constructed primarily from bamboo culms (Dillon, n.d.).

The durability of bamboo as a building material is one of its chief benefits. The hollow culms (stems) of bamboo confer to the plant its characteristic light weight and durability (Nurdiah, 2016). Bamboo can be utilized in load-bearing constructions such as beams and columns due to its greater compressive strength than many forms of wood (Nurdiah, 2016). The great tensile strength of bamboo makes it perfect for use in flooring, scaffolding, and other applications where it must endure tremendous weights (Watts, 2019). Another benefit of bamboo is its rapid growth rate. Bamboo can grow up to 91 centimeters every day, making it one of the world's fastest-growing plants (Xiao, 2008). It reaches maturity in three to five years, but many hardwood trees require decades to attain maturity (Xiao, 2008). This makes bamboo a renewable and sustainable resource that can be grown rapidly and without the use of toxic chemicals. Additionally, bamboo is an economical building material (Neely, 2022). It is readily accessible and may be grown locally, reducing transportation costs and carbon emissions. Moreover, compared to other building materials such as wood and concrete, bamboo is comparatively inexpensive (Scheer, 2005). In addition to its practical applications, bamboo has numerous environmental advantages. It is a carbon sink, meaning it collects carbon dioxide from the atmosphere and can help mitigate climate change (Scheer, 2005).

Bamboo Preservation Techniques and Treatments

Preserving bamboo is an essential aspect of using this renewable resource as a building material. As a natural material, bamboo is susceptible to degradation from factors such as weathering, insects, and fungi. Therefore, bamboo structures require proper preservation techniques to ensure their durability and longevity (Nurdiah, 2016).

Utilizing preservatives such as borates or borax is one of the most prevalent preservation techniques. These preservatives prevent insect damage and rot and also increase the bamboo's fire resistance. The bamboo is treated by dipping it in a borate or borax solution for several hours, followed by air-drying (Nurdiah, 2016). Using a combination of preservatives and fire retardants is yet another method of preservation. This process involves applying a fire-retardant chemical to the surface of the bamboo and then dipping it in a preservative (Neely, 2022). This double treatment increases the bamboo's resistance to fire and protects it from insect damage and decay (Neely, 2022). Using natural oils and waxes is another method of preservation. This method involves applying natural oils, such as linseed oil or tung oil, or waxes, such as beeswax, to the bamboo's surface. These oils and waxes provide a protective layer that prevents the bamboo from absorbing moisture and shields it from insect damage (Scheer, 2005). Heat treatment is another method of bamboo preservation that involves exposing the bamboo to high temperatures to kill any insect larvae or eggs that may be present. This method is particularly useful in regions where bamboo-boring insects are prevalent. The heat treatment procedure is conducted in a kiln where the temperature is maintained at a specific level for several hours (Nurdiah, 2016).

In addition to these preservation techniques, the storage and transportation of bamboo play an important role in its preservation (Neely, 2022). To prevent moisture from entering and causing rot, bamboo should be kept in a well-ventilated, dry area. During transport, bamboo should be shielded from direct sunlight and precipitation to prevent deterioration (Neely, 2022).

Construction Methods for Bamboo Structures

Bamboo is a versatile material that has been used for construction for thousands of years. Due to its strength, durability, and adaptability, bamboo has proven to be an ideal building material for a variety of construction projects, ranging from small structures to large buildings. However, the success of a bamboo structure largely depends on the construction methods used (Scheer, 2005).

The use of split bamboo or bamboo poles is one of the most common and traditional methods for constructing with bamboo. The bamboo culms are cut into thin strips, which are then used as the primary structural members (Neely, 2022). The strips are held together using bamboo joints, bamboo nails, or bamboo ropes. This method is straightforward and economical, and it can be used to build a variety of structures, including houses, bridges, and fences (Neely, 2022).

Using bamboo mats or panels is an additional building method. This method involves weaving bamboo strips into mats, which are then used as the main structural members (Schroder, n.d.). The mats can be used as wall or roof panels, or as flooring. As the mats are easily transported and assembled on-site, this method is particularly useful for the construction of large, open spaces (Nurdiah, 2016).

The use of bamboo scaffolding is another method of construction that dates back centuries. Bamboo scaffolding is a temporary structure used to support workers and building materials during building construction (Neely, 2022). The scaffolding is made of bamboo poles that are tied together using bamboo ropes and is supported by bamboo cross beams (Nurdiah, 2016). This method is cost-effective, lightweight, and can be easily assembled and disassembled (Scheer, 2005).

Bamboo laminates or composites are another construction method that is becoming increasingly popular. This method involves bonding bamboo strips together using adhesives to form a composite material (Dillon, n.d.). The composite material can then be used as the main structural members, or as a cladding material. This technique produces a material that is uniform, sturdy, and suitable for the construction of large structures (Scheer, 2005).

In addition to these construction methods, there are several innovative bamboo-based construction techniques. For example, the use of bamboo cross-lamination, which involves layering bamboo strips at right angles to each other, provides a highly durable material that is suitable for use in walls, floors, and roofs (Janssen (auth.), 1991; Silverman, 2021). Utilizing bamboo prefabrication, in which bamboo components are prefabricated and then transported to the construction site for assembly, is a cost-effective and efficient method for constructing bamboo structures.

There are a variety of construction techniques available for bamboo structures, ranging from traditional techniques such as split bamboo and bamboo scaffolding to innovative techniques such as bamboo composites and prefabrication (Janssen (auth.), 1991). The type of construction method used will depend on several factors, including the location, climate, and intended use of the structure. For the success and durability of bamboo structures, proper design, engineering, and construction practices are required regardless of the method employed (Janssen (auth.), 1991).

Bamboo Reinforcement and Connections

Reinforcement and connections made from bamboo are essential components in the construction of bamboo structures. Reinforcing bamboo increases its tensile strength and torsional rigidity, while proper connections ensure that the structure is securely held together (Nurdiah, 2016).

Utilizing bamboo laminates or composites is one of the most common methods for reinforcing bamboo. This technique involves adhering multiple layers of bamboo strips together to create a composite material with enhanced strength and stability (Nurdiah, 2016). Bamboo laminates can be utilized as reinforcement for the primary structural members or as a cladding material to add strength and stability to the structure (Nurdiah, 2016). Utilizing steel or other metal elements to reinforce bamboo is another option. Steel rods, cables, or plates can be used to reinforce bamboo, giving the structure added strength and stability (Nurdiah, 2016). This technique is especially useful in high-stress areas, such as bridges, or in areas where the structure is exposed to extreme weather. Also essential for the stability and durability of bamboo structures are proper connections. There are a number of ways to connect bamboo elements, such as bamboo joints, bamboo nails, and bamboo ropes (Silverman, 2021). Typically, bamboo joints are created by interlocking two bamboo elements and securing them with bamboo nails or bamboo ropes. Bamboo joints offer a sturdy, secure connection that can withstand substantial stresses and loads (Neely, 2022). Using bamboo dowels is an alternative method for connecting bamboo elements. Bamboo dowels are cylindrical pieces of bamboo used to connect two components by inserting one end into one component and the other end into the second component. This technique provides a sturdy, secure connection that is applicable to a wide variety of bamboo structures (Nurdiah, 2016).

In addition to these methods for reinforcing and connecting bamboo elements, a number of innovations exist in bamboo reinforcement and connections. For instance, bamboo-reinforced concrete, in which bamboo is used to reinforce concrete structures, is a sustainable and cost-effective alternative to conventional reinforcement techniques (Dillon, n.d.). The use of bamboo-reinforced steel, in which bamboo is used to reinforce steel structures, offers a strong and long-lasting reinforcement option that is suitable for use in high-stress areas (Schroder, n.d.).

In conclusion, bamboo reinforcement and connections are essential components of bamboo structure construction. Several techniques exist for reinforcing and connecting bamboo elements, ranging from the use of bamboo laminates and metal elements to innovative techniques like bamboo-reinforced concrete and bamboo-reinforced steel (Neely, 2022). To ensure the success and durability of bamboo structures, proper design, engineering, and construction techniques are essential (Nurdiah, 2016).

Bamboo in Popular Culture

Throughout history, bamboo has played a significant role in popular culture in a variety of regions of the globe. In Asia, for instance, bamboo is frequently connected with longevity, flexibility, and resiliency, and it serves as a symbol for these traits in art, literature, and daily life (Nurdiah, 2016).

Bamboo is one of the Four Gentlemen, four plants that represent the four seasons in Chinese culture. Bamboo symbolizes spring and is associated with vigor, honesty, and longevity. It is also commonly employed in Chinese art, literature, and architecture as a sign of fortune (Schroder, n.d.). In Chinese literature, bamboo is frequently employed as a metaphor for a person's moral character, with its straightness and height indicating integrity and its hollow inside representing humility (The Constructor, 2016).

In Japanese culture, bamboo also represents strength, integrity, and resiliency. It is frequently used in Japanese gardens and a common topic in Japanese art, especially in the shape of bamboo groves. Bamboo flutes are popular instruments in traditional Japanese music, where bamboo is a significant ingredient (Janssen (auth.), 1991).

In Southeast Asian cultures, bamboo is an integral component of daily life, as it is utilized for a variety of uses, including building materials, furniture, and textiles. Bamboo is also an integral part of traditional

Southeast Asian music, since bamboo xylophones and other instruments are frequently seen at festivals and other events (Janssen (auth.), 1991).

In Western culture, bamboo is frequently connected with the tropical and subtropical climates where it grows, and it is frequently used in tropical-themed décor, clothes, and accessories. Bamboo is also a popular material for outdoor furniture and is frequently used to make fences and screens with a natural appearance. Bamboo is also employed as an eco-friendly and sustainable alternative to conventional building materials in architecture (Watts, 2019).

In recent years, bamboo has gained appeal as a sustainable material and has been incorporated into a wide range of products, including clothes, home décor, tableware, and more. Bamboo goods are deemed eco-friendly due to their rapid growth, lack of need for pesticides or fertilizers, and biodegradability (Silverman, 2021).

References

Dillon, V. (n.d.). The Momentum | Bamboo As a Building Material. Retrieved January 31, 2023, from https://www.themomentum.com/articles/bamboo-as-a-building-material

Encyclopedia Britannica. (2022). Concrete | Definition, Composition, Uses, Types, & Facts | Britannica. https://www.britannica.com/technology/concrete-building-material

Hebel, D. (2018, January 31). Natural building materials: Bamboo. RICS. https://www.rics.org/en-in/news-insight/future-of-surveying/sustainability/natural-building-materials-bamboo/

Janssen (auth.), J. J. A. (1991). Mechanical Properties of Bamboo (1st ed.). Springer Netherlands.http://gen.lib.rus.ec/book/index.php?md5=1f8e97dfc9769b5a814eb2787dc5d86d
Neely, A. (2022, March 22). Is Bamboo the Building Material of the Future? NUVO. https://nuvomagazine.com/daily-edit/is-bamboo-the-building-material-of-the-future

Nurdiah, E. A. (2016). The Potential of Bamboo as Building Material in Organic Shaped Buildings. Procedia - Social and Behavioral Sciences, 216, 30–38. https://doi.org/10.1016/j.sbspro.2015.12.004

Scheer, J. (2005). How to Build With Bamboo: 19 Projects You Can Do at Home (First Edition). Gibbs Smith, Publisher. http://gen.lib.rus.ec/book/index.php?md5=58521c4b925e890800ea3a7406be2e56

Schroder, S. (n.d.). Advantages of Building with Bamboo. Guadua Bamboo. Retrieved January 31, 2023, from https://www.guaduabamboo.com/blog/advantages-of-building-with-bamboo

Silverman, E. (2021, December 17). Can Bamboo Construction Material Replace Timber and Steel? | Built. Built | The Bluebeam Blog. https://blog.bluebeam.com/building-with-bamboo/

The Constructor. (2016, December 6). Bamboo as a Building Material—Its Uses and Advantages in Construction. The Constructor. https://theconstructor.org/building/bamboo-as-a-building-material-uses-advantages/14838/

Watts, J. (2019, February 25). Concrete: The most destructive material on Earth. The Guardian. https://www.theguardian.com/cities/2019/feb/25/concrete-the-most-destructive-material-on-earth

Xiao, Y. (2008). Modern Bamboo Structures: Proceedings of the First International Conference (1st ed.). http://gen.lib.rus.ec/book/index.php?md5=f0a2b2ff3e6ae61bf7ea124681ffcbc3

Chapter 10: Popular Projects Using Alternative Materials
by Daniel Gurin

Introduction

There is no shortage of ideas when it comes to alternative materials for building infrastructure. From repurposed waste to ferrock, many sustainable products have been explored as viable replacements for traditional construction materials such as concrete. That being said, there is still much skepticism and reluctance to use these alternatives on a larger scale. By examining some notable projects that use sustainable alternatives as primary construction materials, the true potential of these substances can be emphasized.

Plastic Bottle Roads

One of the many ways an alternative material has been used instead of concrete is using recycled plastics to create roads and highways. Not only does this plastic-infused road material cause less harm to the environment, but it is also a workaround for some of the deterioration facing conventional concrete and asphalt roads (Canadian Plastics, 2021). External factors such as freeze-thaw cycles, UV rays, moisture, and traffic pressure cause the wear of traditional asphalt and concrete. This leaves cracks and potholes for drivers to deal with. Adding recycled plastic into the mix may significantly reduce this deterioration by increasing the durability of the roads (Canadian Plastics, 2021).

Besides the durability-related benefits, using recycled plastic in creating roads also significantly benefits the environment. For one, it increases the efficiency of resources and creates a sense of circular economy by reducing post-consumer plastic going into landfills (Christy, n.d.). The level of plastic used in the creation of these roads varies from project to project. Companies such as Ontario-based GreenMantra Technologies have been leading the way in terms of new ways to incorporate recycled plastic into road construction (Christy, n.d.).

One extreme has been seen in several developing countries, where plastic waste was simply compacted and utilized to pave an entire road with minimal other materials. While not as structurally sound as when different materials are added to the mix, this method is a fantastic alternative to traditional recycling and concrete paving (Canadian Plastics, 2021). Unfortunately, as plastic is far less sturdy than concrete and asphalt, these plastic roads are more adequate for lower traffic conditions and bike and pedestrian pathways. They also do not meet the standards required by transportation authorities in Canada and the United States, barring the method from viable use in North America (Canadian Plastics, 2021).

Use of RPM Asphalt Mixes in Canada

A type of plastic road that's much less extreme and subsequently more feasible is a recycled plastic-modified (RPM) asphalt mixture, where recycled plastics that come primarily from water bottles and single-use plastic bags are turned into plastic pellets. These pellets are then mixed with hot asphalt to form an RPM asphalt binder. For this process to properly function, different plastics must first be sorted by polymer type, reflecting the specific plastic's durability and flexibility (Canadian Plastics, 2021).

The first recorded example of a plastic-asphalt road project in Canada was in the early 1980s. The City of Edmonton ground up children's toys to repave a local street. Later, in 1990, a section of Highway 401 near Toronto, the busiest highway in Canada, was paved using RPM asphalt sourced from recycled milk jugs (Canadian Plastics, 2021). Similar projects popped up here and there throughout the 1990s, most of them using non-recycled plastics, but sparked up in high gear in the late 2000s. With improving technology, RPM asphalts have become more and more prevalent and have been used in a variety of Canadian paving projects (Canadian Plastics, 2021). Vancouver International Airport paved one of its runways using RPM asphalts. Similar projects were done on an expressway along the Fraser River in British Columbia, as well as several parking lots in Nova Scotia, which utilized 2 tonnes of material sourced from plastic shopping bags (Canadian Plastics, 2021).

The aforementioned GreenMantra's first full-length paving project took place in Vancouver in 2014, called the "Blue Box to Green Roads" project. The project let the City of Vancouver use GreenMantra's recycled plastic waxes to pave some of their roads, which helped the environment and allowed for lower paving temperatures during construction (Christy, n.d.). This subsequently generated lower energy demands and lower levels of harmful fume emissions. Another large project led by GreenMantra was one in partnership with Nova Chemicals Corp., where the company used its polyethylene wax to pave two Sarnia pathways. The pathways were about 1,700 feet by 8 feet and about 2 inches thick, which used up about 113,000 single-use plastic bags (Canadian Plastics, 2021). The company is looking forward to other new opportunities and currently has early-stage projects and trials underway in North America, Europe and the Middle East. GreenMentra estimates that each kilometer of road paved with their recycled plastic waxes can repurpose the equivalent of 2.6 million single-use plastic bags, keeping them out of landfills. This would prove to be a massive positive impact on the environment and may inspire more companies and cities to undertake similar projects (Christy, n.d.).

Worldwide Use of RPM Asphalt Mixes

From a worldwide perspective, Europe and other parts of the world are also exploring recycled plastic-based paving projects. India, a pioneer in the widespread mixing of plastics and asphalt, laid down their first waste plastic road in 2002. Since then, the Indian city of Bangalore has paved about 3,000 kilometers of their streets with RPM asphalt mixes (Canadian Plastics, 2021). Dow Chemical Co. has recently collaborated with India's government and waste collectors in Bangalore to pave 40 kilometers of roads with polymer-modified asphalt technology, which redirected about 25 million plastic pouches away from landfills. Similar projects have also been seen in Thailand, the Philippines and Vietnam (Canadian Plastics, 2021).

MacRebur, a Scottish asphalt specialist and international supplier, offers a diverse line of RPM pellets used for asphalt paving. Each kilometer of road laid with the pellet mix for these specific pellets utilizes about 740,541 single-use plastic bags. The company has led paving projects across the U.K., as well as in New Zealand, Australia, and South Africa (Canadian Plastics, 2021).

In July 2020, a paving company in California repaved three lanes on a 1,000-foot section of an Oroville highway using recycled asphalt pavement and melted plastic from single-use bottles. This marked the first time the California Department of Transportation used 100% recycled material to pave a road (Canadian Plastics, 2021). In another California-based project, TechniSoil Industrial used a "recycling train of equipment" that tears apart the top three inches of pavement, mixes the grindings with a liquid plastic polymer binder, and then lays the new asphalt material back on the surface of the road in one fell swoop (Canadian Plastics, 2021). The plastic polymer used is sourced from a high amount of recycled single-use bottles. This project repurposes plastic sets for a landfill, provides an easier alternative to traditionally repaving damaged roads, and eliminates the need for trucks to bring in external materials for paving operations (Christy, n.d.).

As of early 2021, there were about 200 complete and ongoing projects throughout the world. As time progresses, the durability of these projects will come to light and may offer suggestions for the future of the recycled plastic-asphalt industry. Regardless, the industry is making meaningful strides toward creating a more sustainable world and is helping redirect millions of plastic items away from landfills and into much-needed infrastructure (Canadian Plastics, 2021).

Scotland's Plastic Bridge

Scotland has long been a leader in sustainability initiatives, with a full-scale prohibition of single-use plastics such as drinking straws and bags since the end of 2019. The River Tweed in the Scottish county of Peeblesshire is now home to the Easter Dawyck Bridge, which was built using about 50 tonnes of 100% recycled materials and can hold up to 44 tonnes of vehicular weight (Iberdrola, 2011). The bridge was constructed in 2011 by local contractors in under three weeks and was made to resemble bridges traditionally built in the county. The recycled plastic material created a light and highly stable bridge that could not corrode nor rust and would require minimal maintenance (Iberdrola, 2011).

The Easter Dawyck Bridge project was designed and developed by Vertech Ltd with the help of the Welsh Regional Government, alongside the Engineering School of Cardiff University and Rutgers University. Rutgers University was responsible for developing the thermoplastic material known as structural plastic lumber. One of its researchers, Thomas Nosker, saw an opportunity in the many disadvantages of plastic: he saw the potential for long-lasting building materials with a premise of continuous recyclability (Iberdrola, 2011).

Nosker's team initially set out to explore high-density polyethylene, a material commonly found in industrial packaging, to replace treated wood used in the construction of small-scale items such as garden benches. The project ran into trouble as the recycled polyethylene planks were too weak to withstand constant use and broke quickly (Iberdrola, 2011). However, after that, the team had a breakthrough when they combined food takeout box polystyrene and more rigid plastics in various proportions. The resultant material proved perfect for constructing buildings due to its profound structural integrity and durability compared to wood. It also required essentially zero maintenance and, most importantly, was 100% recycled and recyclable (Iberdrola, 2011).

Another critical benefit of Nosker's material is that this combination of rigid plastics is very pliable when hot, meaning that it can take on any shape and form, opening a plethora of opportunities for civil engineering projects (Iberdrola, 2011). The material could be used to replace concrete and other building materials for projects such as railway ties, bridge structures, and various structural columns, such as those supporting above-ground highways (Iberdrola, 2011).

Nosker currently has a partnership with a private United States company which has resulted in various patents that have gotten the material up to standard in terms of technical specifications, which has since then reduced the costs associated with the material (Iberdrola, 2011).

New York's Museum of Modern Art Mushroom Wall

Another intriguing example of alternative material to concrete is mycelium - the underground, vegetative portion of a fungus. Mycelium consists of high-density filaments of hyphae and is crucial to the life-support system for the fungus (Knightsbridge, 2021). On top of its importance to the sustenance of fungal life, something curious happens when mycelium is mixed with agricultural waste. When mycelium is added to a mixture along with cut-up straw or corn husks and then placed into a brick mold for several days, the result is a firm yet lightweight brick (Knightsbridge, 2021).

Grown without any carbon emissions or waste, these bricks are incredibly sustainable and pose as a viable replacement for real bricks and concrete. They are, however, often mixed with clay and even cement to support their integrity and durability, but always in minimal amounts. Their use of agricultural waste also serves as an effective and eco-friendly way to reuse otherwise useless materials (Knightsbridge, 2021).

In 2014, architect David Benjamin's design of a tower built from 100% mycelium bricks was erected at the Museum of Modern Art in New York. The tower used over 10,000 of the aforementioned mycelium bricks and stands over 12 meters tall in one of the museum's courtyards. The bricks took five days to grow into their shapes, and the additives used in the mycelium mix were sourced from local biodegradable waste, including rice hulls (Arthur, 2014).

The imaginative concept of David Benjamin revolved around the idea of a "living factory," where living systems essentially grew the building materials used. The mushroom tower is a part of Benjamin's mission to explore the notion of biotechnology and is just one project led by his appropriately-named architectural firm, The Living (Frearson, 2014).

Manav Sadhna in India's Ahmedabad

Filled with school grounds, vocational training centers, gymnasiums, arts and crafts centers and more, the Manav Sadhna Centre in Ahmedabad, India, is a fantastic example of sustainable construction (Rethinking The Future, 2021). Its building components feature almost exclusively recycled waste. The 1,100 square meter project, led by architect Yatin Pandya, was a public demonstration against pollution and the wasting of energy.

The project used waste such as fly ash, landfill waste, packing from various types of crates, single-use plastic water bottles, various glass containers, broken ceramic wares, used electronic hardware, and more (Rethinking The Future, 2019). These materials were repurposed into different types of walling, roofing, flooring, and insulation throughout the center. Producing roughly 2,750 metric tonnes of waste daily, the city of Ahmedabad had plenty of waste material to spare for Pandya's ambitions.

The project's goal was to combine the demonstration of different applications of recycled waste with low-cost and aesthetically pleasing building materials (Rethinking The Future, 2019). The process of repurposing local waste incorporated preparation using hand-operating tools with the assistance of locals - many of whom were to be end-users of the center.

The main components of the center's walls were cement-bonded fly ash bricks, often known as Ashcrete (Rethinking The Future, 2019). They also used mold-compressed bricks consisting of landfill waste, strengthened soil blocks, recycled glass and plastic bottles solidified by being filled with ash and waste residue, and various sheets of wood from crates. Regarding roofing and flooring, filler slabs were made with a combination of glass bottles, plastic bottles, bricks, stone slabs, old piping, clay, and other recycled materials.

To adhere to the aesthetic goals of Pandya's project, the slabs and other products used as building blocks were developed and produced prior to the construction of the building (Rethinking The Future, 2019). They also underwent rigorous testing to confirm strength and durability to ensure the safety of the end-users of the center. All in all, this project perfectly blended innovation, environmental consciousness, and aesthetic design into one multi-purpose public center.

The Wheel Story House, Ghana

Designed by one of the greatest and most respected architects in West Africa, Samuel Mensah Ansah, the Wheel Story House is the largest and oldest building in West Africa that is made from recycled materials (Ntim, 2020). The project was Ansah's stand against plastic waste pollution and sourced local waste from the Tema Port, decommissioned gas stations and discarded electronics.

The house has served as a residential building containing 12 apartments and was constructed out of 100% reclaimed and reused materials (Ntim, 2020). Some of the most common materials found in the project include broken coffee cups, used stones and timber, abandoned shipping containers and discarded electrical wiring. Ansah hoped the project would serve as an artistic statement and help visitors understand that nothing should be considered "waste." He also hoped that people, after witnessing his creation, would feel inspired to construct their own recycled-trash creations.

Built in 2005, the residence's 18-year life so far is a testament to the potential durability of recycled materials and the viability of using them for residential and commercial buildings (Ntim, 2020). Not only is it fantastic for the environment as otherwise landfill-set trash is repurposed as building materials, but builders can also significantly reduce costs, making these types of projects perfect solutions for housing in poverty-stricken areas.

Redondo Beach House, California's Cargo Container Getaway

Another intriguing example of recycling abandoned materials to construct living spaces is the recent interest in reusing cargo containers. One prime example of this is the Redondo Beach Shipping Container House, designed and built by Peter DeMaria's architectural firm (Eco Container Home, n.d.). DeMaria's design is as visually stunning as it is eco-friendly, with a modern yet cozy atmosphere.

The cargo containers used were adjusted to the requirements of the design by cutting out holes and trimming walls (Eco Container Home, n.d.). The resulting "building blocks" were then attached to one another to form the house using welding and minimal traditional building materials such as cement. Not only are these cargo containers extremely durable, but they are also fire-proof and termite-proof and come with many other benefits. Their properties make the final product a well-built, exceptionally environmentally friendly home.

Reusing cargo containers also saves a lot of money, as there is no need for structural support beyond the thick steel walls of the containers themselves (Eco Container Home, n.d.). This makes them ideal solutions to creating living spaces in places of poverty, but just like the Redondo Beach House, they can also prove to be great statements of art combined with eco-friendliness.

Keetwonen, Amsterdam's Cargo Container Student Housing Complex

Taking the idea of reusing decommissioned cargo containers as building materials to the extreme, Amsterdam company TempoHousing built a multi-storey student residence complex using over 1,000 cargo containers (Martinez, 2017). The project only took eight months to complete, thanks to China's swift 40-unit per week production and was ready for use in 2006.

The complex quickly became one of Amsterdam's most popular student housing locations (Martinez, 2017). It attracted students not just for the uniqueness of the design but also for its excellent amenities. The complex hosted a wide array of features, from laundry mats to cafes and common areas, making it no worse than traditional student living centers.

The cheap cost of production, thanks to the re-use of the abandoned shipping containers, allowed for very affordable rent for students, with the 2015 rate hovering around $475 USD per month (Martinez, 2017). The durability of the containers also proved very reliable, even extending the anticipated project life of the complex. The complex was initially supposed to be taken down after five years of use, but due to the lack of deterioration, it was extended to 2018.

Hotel Magdalena, Austin's Mass Timber Hotel

Mass timber is among the most promising alternative building materials, as it has the potential to replace concrete entirely (Picó, 2021). Mass timber refers to several products, including veneer lumber, nail-laminated wood, glue-laminated beams, and other alternatives. Arguably, the best of these alternatives is the cross-laminated timber (CLT) variant, where solid planks of timber are cut into shape, dried and then glued together under high pressure and in weaving directions. The resultant wood panels are incredibly strong and serve as durable building materials.

Using mass timber as a primary construction material, architect team Lake Flato, with the help of engineer-builders StructureCraft, constructed North America's first hotel made from mass timber (StructureCraft, n.d.). The Magdalena Hotel in Austin, Texas, features a set of three buildings of varying heights and uses dowel-laminated timber (DLT) mass timber panels for the majority of its structure. The design successfully leveraged the material's properties to create a visually appealing and structurally sound building.

The wood-based material has significantly lower carbon emissions compared to traditional building materials (StructureCraft, n.d.).

Lake Flato and StructureCraft ran a baseline design to demonstrate the environmental difference made by building the hotel using mass timber. The result was a 38% reduction in global warming potential compared to the baseline. While small batches of concrete were used where absolutely necessary, the bulk of the hotel, including the support beams and both interiors and exteriors, were made of DLT mass timber and glulam beams. Glulam beams are engineered wood designed to be pound-for-pound stronger, and firmer than steel (APA, n.d.).

Conclusion

The examples explored in this chapter are only a glimpse at the limitless possibilities offered by alternative materials. The foremost reason for using alternative materials over traditional building materials such as concrete remains the need to be environmentally conscious and to build sustainably. The examples highlighted prove that structural integrity and durability can still be attained without the excessive use of concrete and other environmentally damaging materials.

References

APA. (n.d.). Glulam—APA – The Engineered Wood Association. Retrieved January 30, 2023, from https://www.apawood.org/glulam

Arthur, G. (2014, August 29). Making houses out of mushrooms. BBC News. https://www.bbc.com/news/magazine-28712940

Canadian Plastics. (2021, February 5). The plastic road ahead. Canadian Plastics. https://www.canplastics.com/features/the-plastic-road-ahead/

Christy. (n.d.). ASPHALT. GreenMantra Technologies. Retrieved January 16, 2023, from https://greenmantra.com/asphalt-applications/

Conserve Energy Future. (2020, September 11). 17+ Sustainable and Green Building Construction Materials—Conserve Energy Future. https://www.conserve-energy-future.com/sustainable-construction-materials.php

Eco Container Home. (n.d.). The Redondo Beach House Container Home—Eco Container Home—Shipping Container Homes, Cargo Homes & Green Building. Retrieved January 30, 2023, from https://ecocontainerhome.com/redondo-beach-house/

Expert-Market. (2020, May 6). Don't Like the Idea of Using Concrete? Here Are Ten Natural Alternatives | Expert-Market. https://www.expert-market.com/dont-like-the-idea-of-using-concrete-here-are-ten-natural-alternatives/

Frearson, A. (2014, July 1). Tower of "grown" bio-bricks by The Living opens at MoMA PS1. Dezeen. https://www.dezeen.com/2014/07/01/tower-of-grown-bio-bricks-by-the-living-opens-at-moma-ps1-gallery/

Iberdrola. (2011). El puente de plástico reciclado más largo del mundo, en Escocia. Iberdrola. https://www.iberdrola.com/sustainability/scotland-home-worlds-longest-recycled-plastic-bridge

Knightsbridge, A. (2021, July 28). 5 Eco Friendly Alternatives To Concrete. https://knightsbridgecorp.ca/5-eco-friendly-alternatives-to-concrete/

Martinez, O. (2017, October 2). Keetwonen, Amsterdam's 1000 Cargo Container Student Housing Complex. https://www.containerpedia.com/keetwonen-amsterdam-s-1000-cargo-container-student-housing-complex

Morrison, R. (2021, June 30). Alternatives to Concrete and Concrete Blockwork—Bricsys Blog. https://www.bricsys.com/en-ca/blog/alternatives-to-concrete-in-construction

Ntim, A. (2020, August 28). The Wheel Story House: A home made from recycled materials. MyGhanaDaily. http://myghanadaily.com/the-wheel-story-house-a-home-made-from-recycled-materials/

Picó, R. (2021, July 14). 9 Mass Timber Projects Inspiring Change in the Industry. Gb&d Magazine. https://gbdmagazine.com/mass-timber-projects/

Rethinking The Future. (2019, June 9). Manav Sadhna by Yatin Pandya. RTF | Rethinking The Future. https://www.re-thinkingthefuture.com/architecture/housing/manav-sadhna-by-yatin-pandya/

Rethinking The Future. (2021, May 7). 15 architectural projects made out of recycled materials. RTF | Rethinking The Future. https://www.re-thinkingthefuture.com/designing-for-typologies/a4102-15-architectural-projects-made-out-of-recycled-materials/

StructureCraft. (n.d.). Magdalena Hotel in Austin Texas | Timber Boutique Hotel. Retrieved January 30, 2023, from https://structurecraft.com/projects/magdalena-hotel

Chapter 11: The Timeline of Concrete
by Lea Touliopoulos

Throughout most of the 21st century, there are many things that we take to be a regular aspect of our lives without considering where they come from or what life would be like without them. One example of such a thing is concrete. In our current day and age, concrete is everywhere and used for a multitude of purposes, from roads, to construction of buildings and more. However, concrete is not something that many people stop to think about or even know where it came from. There are many questions that could be asked concerning concrete. For example, when was concrete first invented? Who first invented concrete? How much did this preliminary form of concrete resemble what is used today? It would be fair to conclude that most people do not know the answers to these questions, and are not educated on the history and subject of concrete.

Starting with the first question, "When was concrete invented?". This question unfortunately does not have a simple answer, partly due to ambiguities in how the term "concrete" can and should be interpreted. For example, there are records of there being ancient materials that existed that resembled crude cements (Gromicko & Shepard, n.d) . These cements were made by the crushing as well as the burning of gypsum or limestone (Gromicko & Shepard, n.d). Once this step had been completed, sand and water were thought to have been added to the mixture, which gave the mixture a thick mortar like texture (Gromicko & Shepard, n.d). This mortar was then used to stick stones to each other (Gromicko & Shepard, n.d). Evidently, this is very different from what is used as concrete today, but the resemblance between the two can still be seen.

What many consider to be the first significant precursor to concrete was invented in 1300 BC. This early form of concrete was found when builders located in the Middle East realized that it was possible to coat their buildings with a thick, damp coating of burned limestone (Gromicko & Shepard, n.d). The reason that doing this was advantageous was because this burned limestone would be able to react chemically with gasses that were present in the air (Gromicko & Shepard, n.d). This causes a hard, protective surface to form. It is not hard to try and imagine the uses that the residents of the Middle East might have had for such a material. One of the most obvious advantages would be protection from the weather elements, as well as for shelter and defence (Gromicko & Shepard, n.d). It was soon found that this mix of materials was very common for the initial concrete mixtures that were present centuries ago (Gromicko & Shepard, n.d). A typical early concrete mixture would have contained mortar-crushed, burned limestone, as well as sand and water (Gromicko & Shepard, n.d).

However, even before humans began refining their abilities of producing and using concrete. cement has been around in natural deposits (Gromicko & Shepard, n.d). To give some scope of how old these natural deposits are, there was one that was known to have formed around 12 million years ago (Gromicko & Shepard, n.d). One question that must be addressed is what is a natural deposit and how does concrete form in them. The answer to this is natural deposits were formed by reactions between limestone and oil shale (Gromicko & Shepard, n.d). The reason this reaction was able to move towards completion is because it is a spontaneous corruption (Gromicko & Shepard, n.d).

Another interesting aspect is who first discovered that concrete could be used to build structures. The first known concrete-like structures were built by the Nabataea traders or the Bedouins (Gromicko & Shepard, n.d). In 6500 BC, these groups were able to occupy and control a series of oases and develop a small empire in the regions of southern Syria and northern Jordan (Gromicko & Shepard, n.d). The Nabataea had a different perspective on concrete. The approach that the Nabataea took was making the mix they used as concrete as dry and low-slump as possible (Gromicko & Shepard, n.d). Their reasoning for this was that using excess water in the mixture would introduce a weakness into the concrete (Gromicko & Shepard, n.d). The Nabatara were also thought to use a process called tamping (Gromicko & Shepard, n.d). They would do this after the concrete had been freshly placed and they would use special tools (Gromicko & Shepard, n.d). What is interesting about this process is that it would produce gel, or in other words, the bonding material produced by chemical reactions that occur during hydration (Gromicko & Shepard, n.d). It is important to note that this tamping process binds the particulates and the aggregate together (Gromicko & Shepard, n.d). This process allowed there to start to be homes being built using this mixture, which resembles concrete, for a floor (Gromicko & Shepard, n.d).

Next moving on to Egypt, where a different technique for making a concrete mixture is found. Around the year 3000 BC, it is thought that the ancient Egyptians developed the technique of mixing mud and straw together (Gromicko & Shepard, n.d). They would then use this mixture to form bricks (Gromicko & Shepard, n.d). While this mixture resembles modern day concrete less than the other mixtures that were discussed earlier, the ancient Egyptians also did use a more typical concrete like mixture that contained gypsum and lime mortars (Gromicko & Shepard, n.d). It was this mixture that the Egyptians used to build the pyramids, many of which are still standing today and are considered a wonder of the ancient world (Gromicko & Shepard, n.d).

However, the ancient Egyptians were not the only ones who were making progress working with concrete. Also around the year 3000 BC the northern Chinese were also constructing what would become a world famous structure (Gromicko & Shepard, n.d). The Great Wall of China was being built around this time, but the mixture of ingredients the ancient Chinese developed differed in composition from what the result of the world was using (Gromicko & Shepard, n.d). Later, once spectrometer testing was available, it was revealed that the ancient Chinese concrete actually used glutenous, sticky rice as the key ingredient in their concrete (Gromicko & Shepard, n.d). It is also good to note that this concrete mixture was also used as a form of cement during boat-building (Gromicko & Shepard, n.d).

The next notable event that happened in the history of concrete occurred in 600 BC. Moving back to ancient Europe, the ancient Greeks had been able to discover a natural pozzolan material (Gromicko & Shepard, n.d). What was useful about this natural pozzolan material was that it had hydraulic properties (able to move liquids) after it was mixed with lime (Gromicko & Shepard, n.d). While this was the beginning of structures being built using concrete in Europe, the ancient Greeks were never able to reach the mastery and skill of building with concrete that the ancient Romans reached (Gromicko & Shepard, n.d). The ancient Romans' mastery of concrete occurred around 400 years later, by the year 200 BC (Gromicko & Shepard, n.d). However, this fact does not equate to the ancient Romans using the same concrete mixture that we use today to build (Gromicko & Shepard, n.d). Instead the Roman form of concrete could be described more as a cemented rubble (Gromicko & Shepard, n.d). The reason the word rubble can be used to describe this concrete mixture is because the ancient Romans would build by stacking many different stones of many different sizes and then would hand fill the spaces in between them with the mortar like substance (Gromicko & Shepard, n.d). What is ironic is that while the ancient Romans were some of the most skilled and prolific builders with concrete, the concrete did not undergo a true chemical hydration meaning that it was much weaker than the concrete many other civilizations were using (Gromicko & Shepard, n.d).

However, the ancient Romans did have several tricks that they used to make many of their grand and artistic structures. For example, one technique that they used was making a cement mixture from a volcanic sand that is naturally reactive (called Harena Fossicia) (Gromicko & Shepard, n.d). The volcanic sand concrete mixture was used in land-based buildings due to its durability (Gromicko & Shepard, n.d). However, the ancient Romans were also able to build marine structures exposed to salt water and structures exposed to fresh water (example: bridges, docks, storm drains, aqueducts) (Gromicko & Shepard, n.d). For structures that were supposed to be durable near water, it was found that pozzuolana, a different type of volcanic sand, was used instead (Gromicko & Shepard, n.d). It is with these two mixtures that the ancient Romans were able to build many structures that still stand today, such as the Pantheon and the Colosseum (Gromicko & Shepard, n.d).

Furthermore, the ancient Romans took it one step further and were able to manufacture artificial pozzolans as well (Gromicko & Shepard, n.d). These artificial pozzolans were calcined kaolinitic clay and calcined volcanic stones (Gromicko & Shepard, n.d). This is an example of the technology that is being used to make and use concrete starting to slowly advance and become more sophisticated (Gromicko & Shepard, n.d).

While the history of concrete might not be something that many people frequently reflect and think about, clearly it has a long and complex history that spans the globe. From the first records of a concrete-like mixture being developed in the Middle East thousands of years ago, to the Nabataea tribe starting to use the concrete mixture as a floor on their buildings, to the ancient Chinese and the ancient Egyptians using different concrete mixtures to build amazing wonders such as the Great pyramid of Giza and the Great Wall of China concrete has been used in many different ways throughout history. Finally the ancient Greeks and Romans also left their mark on the development of concrete, with the Romans even developing a method of making artificial concrete mixtures.

However, what is modern day concrete like? How is it manufactured? What has changed since ancient times with respect to the concrete industry? How do societies around the world use concrete in the present day? These are just some examples of questions that many people do not think to question, due to concrete being so readily found in our current societies.

Since this paper has already taken the opportunity to discuss how concrete has been made in the past, it is time to move on to how concrete is made in the present. Concrete is made of both fine and coarse aggregate which is mixed together with fluid cement (Gagg, 2014). This mixture can be classified as a composite material, which is defined as a material that consists of at least two different materials that will remain distinct from each other even after the substance is completely mixed (Gagg, 2014). While concrete will start as a liquid, it will harden over time, which is otherwise known as "curing" (Gagg, 2014). The process of making concrete starts by mixing the coarse aggregate with dry cement and water (Li, 2011). This mixture will then form a slurry-like fluid that can be easily poured and moulded into different shapes, contributing to the versatility of concrete (Li, 2011). However, how does concrete go from a liquid slurry to the hard, durable substances that we know as concrete? What occurs is a reaction called concrete hydration, where the cement will react with the water (Li, 2011). This process also causes a hard matrix to form, which will allow the materials to bind together into the hard material that is well known as concrete (Li, 2011). Now to go into some more details about the concrete hydration process. The concrete hydration process is an exothermic process (Industrial Resource Council, 2018). This means that it will release heat into the environment, and that the ambient temperature will have an effect on how long it takes the concrete to set from a liquid slurry to the hard, durable end product (Industrial Resource Council, 2018). On top of the aggregate, the cement, and the water, additives will also often be added (Industrial Resource Council, 2018). The reason additives are added is to attempt to improve the physical properties of the mixture, or to slow down or speed up the curing time (Industrial Resource Council, 2018). Some examples of additives are pozzolans or superplasticizers (Industrial Resource Council, 2018). On top of these additives, the concrete will also usually be poured with a reinforcing material (Industrial Resource Council, 2018). An example of a reinforcing material would be rebar (Industrial Resource Council, 2018). The reason that a reinforcing material is used is to provide tensile strength (Industrial Resource Council, 2018). Tensile strength is what is known as the maximum amount of stress or force that a material can withstand. Adding these reinforcing materials will yield what is known as reinforced concrete, which is able to withstand more force and stress (Industrial Resource Council, 2018). While this is not done as often anymore, lime base cement binders also used to be used in concrete, especially with water resistant cements, known as hydraulic cements (Industrial Resource Council, 2018).

There is also a significant amount of variability in different types of concrete. To elaborate, there are many types of non-cementitious types of concrete that use different methods to bind the aggregate together (Allen & Iano, 2013). For example, there is asphalt concrete, which uses a bitumen binder (Allen & Iano, 2013). This is unsurprisingly often used when paving roads, which is why it is known that roads are made of asphalt. There are also a variety of polymer concrete types that are aptly named as they use different polymers as binders (Allen & Iano, 2013).

An interesting fact about concrete is that it is actually the second-most-used substance in the world, second only to water (Gagg, 2014). This means there is also a huge market for concrete. The largest segment of the concrete market is the ready mix concrete industry, which is expected to have a revenue of more than $600 billion by the year 2025. (Gagg, 2014)

However, the making of concrete is not an environmentally neutral process. One of the principal environmental impacts from concrete production is that it causes the production of large volumes of greenhouse gas (Lehne & Preston, 2018). In fact, enough greenhouse gases are released that the production of concrete contributes to a net 8% of global emissions (Lehne & Preston, 2018). Despite this being such a high number, there are also other environmental concerns surrounding the production and use of concrete (Lehne & Preston, 2018). For example, concrete production has caused widespread illegal sand mining to occur (Lehne & Preston, 2018). Illegal sand mining is when sand is taken from a beach or natural sandy area to be used in industrial production, especially in the production of concrete (Lehne & Preston, 2018). Concrete production and use also causes negative impacts on the surrounding environment. For example, it can cause the urban heat island effect, which is where an urban area is significantly hotter than the surrounding rural areas (Lehne & Preston, 2018). Concrete use also causes issues with surface runoff (Lehne & Preston, 2018). Surface runoff is when the flow of water from rainfall can no longer successfully infiltrate the soil. Concrete use can increase surface runoff as it covers large areas of ground with concrete, making them impermeable to water. Furthermore, concrete production and use can also use toxic ingredients which can have negative public health implications (Lehne & Preston, 2018).

Now that the concept has been introduced, it is time to delve a little further into the environmental impact of concrete. In order to understand where the majority of the greenhouse gas emissions from concrete come from, it is important to understand the composition of concrete. As already discussed, one of the major components of concrete is cement (Akerman et al., 2020). Cement can be described as powder like, and a fine soft substance (Akerman et al., 2020). Its purpose is to hold the fine sand as well as the coarse aggregates together in concrete (Akerman et al., 2020). The way the cement is produced is by mixing a substance called clinker with other additives (Akerman et al., 2020). The additives are always mixed in in lesser quantities than the clinker, and some examples of additives are gypsum and ground limestone (Akerman et al., 2020). Clicker is the material that is present in the highest percentage in cement and it is responsible for the majority of concrete's greenhouse gas emissions (Akerman et al., 2020). The greenhouse gas emissions emitted due to clinker are both energy intensity emissions and process emissions (Akerman et al., 2020).

However, what are the numbers involved in the production of cement and the emission of greenhouse gases? For every tonne of cement produced, how much carbon dioxide is released into the atmosphere? The data shows that on average, a tonne of carbon dioxide is released into the atmosphere for every tonne of cement produced (HeidelbergCement, 2020). This means that carbon dioxide emissions and cement production are currently at a one to one ratio (HeidelbergCement, 2020). However, there does seem to be the possibility that this will improve in the future, as there are new cement manufactures which claim that they can produce cement and emit less carbon dioxide. These new cement manufacturers have data that says they only release 590 kg of carbon dioxide for each tonne of cement produced (HeidelbergCement, 2020). This is evidently a much lower number of carbon dioxide and a step in the right direction for the environment. The reason behind the emissions from concrete are due to the combustion and calcination processes (HeidelbergCement, 2020). The combustion process is thought to be responsible for about 40% of concrete production's greenhouse gas emissions and the calcination process is thought to be responsible for about 60% of concrete production's greenhouse gas emissions (AGICO Cement Plant Supplier, 2019). The fact that concrete is responsible for such a large quantity of greenhouse gas emissions is even more critical when it is considered that more than 10 billion tonnes of concrete are used across the world (Lehne & Preston, 2018). Considering that this is a trend that is not looking to decrease any time soon as concrete is still used in large quantities around the world, it will continue to be increasingly important to try and decrease carbon dioxide and greenhouse gas emission from concrete production.

Evidently, there is a lot more that has gone into concrete than many people consider. Despite it being so prevalent in our society and used in a multitude of different ways, most people have never had the opportunity to learn more about the history, production or use of concrete. Evidently concrete has been around for longer than many people think, and has a history that goes back thousands of years and spans across many civilizations around the globe. Furthermore, it is evident that modern day production of concrete is a process that is both complicated and flawed. Concrete involves many different materials and additives, including cement which is responsible for producing a large amount of greenhouse gas and carbon dioxide emissions. This will continue to be increasingly important as it becomes more and more critical for humans to take steps to address our changing climate and gut back on carbon dioxide and greenhouse gas emissions. However, there is already some work being done on producing carbon in a way that is less harmful to the environment which is a step in the right direction. Overall, it is clear that concrete is incredibly important to our society and it has been incredibly important to humans for many years. While it is not perfect, it is important to understand its history and production in order to improve this product for the future.

References

Akerman, Patrick; Cazzola, Pierpaolo; Christiansen, Emma Skov; Heusden, Renée Van; Iperen, Joanna Kolomanska-van; Christensen, Johannah; Crone, Kilian; Dawe, Keith; Smedt, Guillaume De; Keynes, Alex; Laporte, Anaïs; Gonsolin, Florie; Mensink, Marko; Hebebrand, Charlotte; Hoenig, Volker; Malins, Chris; Neuenhahn, Thomas; Pyc, Ireneusz; Purvis, Andrew; Saygin, Deger; Xiao, Carol; Yang, Yufeng (1 September 2020). "Reaching Zero with Renewables"

Allen, E., Iano, J., (2013). Fundamentals of building construction : materials and methods (Sixth ed.). Hoboken: John Wiley & Sons. p. 314.

"Cement Clinker Calcination in Cement Production Process". AGICO Cement Plant Supplier. 4 April 2019.

Gagg, Colin R. (1 May 2014). "Cement and concrete as an engineering material: An historic appraisal and case study analysis". Engineering Failure Analysis. 40: 114–140. doi:10.1016/j.engfailanal.2014.02.004.

Gromicko, N., Shepard, K., The History of Concrete. (n.d.). Retrieved February 3, 2023, from https://www.nachi.org/history-of-concrete.htm

Industrial Resources Council (2008). "Portland Cement Concrete". www.industrialresourcescouncil.org. Retrieved 20 January 2023.

"Leading the way to carbon neutrality" (PDF). HeidelbergCement. 24 September 2020. Archived (PDF) from the original on 9 October 2022.

Lehne, Johanna; Preston, Felix (2018). Making Concrete Change: Innovation in Low-carbon Cement and Concrete (PDF). London: Chatham House. Archived (PDF) from the original on 9 October 2022.

Li, Zongjin (2011). Advanced concrete technology. John Wiley & Sons. https://onlinelibrary.wiley.com/doi/book/10.1002/9780470950067

Conclusion

The creation of concrete is one of mankind's greatest innovations, and while there can be no denying that concrete has been an essential building material that has played a major role throughout history, its environmental impact cannot be ignored . As we face the pressing challenge of climate change, it is more important than ever to seek out alternatives that offer a greener future for construction. From innovative materials like Ferrock, Ashcrete, and Mycelium, to the rapidly growing popularity of bamboo as a building material, there are a multitude of alternatives to concrete that can help us build a more sustainable future. By embracing these alternatives, and by optimizing the production and disposal of concrete, we can reduce its impact on the environment and help to build a greener, more sustainable world for future generations. With the increasing human population and the strain it places on our resources, it is crucial that we embrace innovative materials and technologies that offer a greener future. The advancements of environmentally friendly options such as the ones explained in detail throughout this book are just a select few examples of the exciting world of green concrete. The future of construction is exciting, and the possibilities are endless. Let's build a better world, one brick at a time.

Conclusion 131

www.ingramcontent.com/pod-product-compliance
Lightning Source LLC
Chambersburg PA
CBHW021914180426
43198CB00035B/660